THE ISLAND OF KNOWLEDGE

ALSO BY MARCELO GLEISER

A Tear at the Edge of Creation

The Prophet and the Astronomer

The Dancing Universe

THE ISLAND
OF KNOWLEDGE

The Limits of Science
and the Search for Meaning

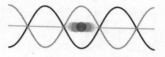

Marcelo Gleiser

BASIC BOOKS
A MEMBER OF THE PERSEUS BOOKS GROUP
New York

Book design by Cynthia Young

Library of Congress Cataloging-in-Publication Data
Gleiser, Marcelo, author.
The island of knowledge : the limits of science and
the search for meaning / Marcelo Gleiser.
pages cm
Includes bibliographical references and index.
ISBN 978-0-465-03171-9 (hardback) — ISBN 978-0-465-08073-1
(e-book) 1. Science—Philosophy. 2. Meaning (Philosophy)
I. Title.

Q175.32.M43G57 2014

501—dc23

2013046890

10 9 8 7 6 5 4 3 2 1

To Andrew, Eric, Tali, Lucian, and Gabriel;

the lights that brighten my path

Contents

PART II

From Alchemy to the Quantum:
The Elusive Nature of Reality

PART III

Mind and Meaning

THE ISLAND
OF KNOWLEDGE

What I see in Nature is a magnificent structure that we can comprehend only very imperfectly, and that must fill a thinking person with a feeling of humility.

—ALBERT EINSTEIN

What we observe is not Nature itself but Nature exposed to our method of questioning.

—WERNER HEISENBERG

How much can we know of the world? Can we know *everything*? Or are there fundamental limits to how much science can explain? If there are limits, to what extent can we understand the nature of physical reality? These questions and their surprising consequences are the focus of this book, an exploration of how we make sense of the Universe and of ourselves.

What we see of the world is only a sliver of what's "out there." There is much that is invisible to the eye, even when we augment our sensorial perception with telescopes, microscopes, and other tools of exploration. Like our senses, every instrument has a range. Because much of Nature remains hidden from us, our view of the world is based only on the fraction of reality that we can measure and analyze. Science, as our narrative describing what we see and what we conjecture exists in the natural world, is thus necessarily limited, telling only part of the story. And what about the other part, the one beyond reach? From our past successes we are confident that, in time, part of what is currently hidden will be incorporated into the scientific narrative, unknowns that will become knowns. But as I will argue in this book, other parts will remain hidden, unknowables that are unavoidable, even if what is unknowable in one age may not be in the next one. We strive toward knowledge, always more knowledge, but must understand that we are, and will remain, surrounded by mystery.

This view is neither antiscientific nor defeatist. It is also not a proposal to succumb to religious obscurantism. Quite the contrary, it is the flirting with this mystery, the urge to go beyond the boundaries of the known, that feeds our creative impulse, that makes us want to know more.

The map of what we call reality is an ever-shifting mosaic of ideas. We will follow this mosaic through the history of Western thought, retracing the steps of our shifting scientific worldview from past to present in three separate but complementary parts. In each part, I strive to illuminate a variety of scientific and philosophical viewpoints, always with the intention of exploring how conceptual shifts inform our search for knowledge and meaning. In Part 1 we will focus on the Universe, its origin and physical nature, and the ways in which our evolving cosmic narrative has framed our understanding of ourselves and of the nature of space, time, and energy. In Part 2 we will focus on the nature of matter and the material composition of the world, from ancient alchemical musings to the most modern ideas about the quantum world and what they tell us about the essence of physical reality and our role in defining it. In Part 3 we explore the world of mind, computers, and mathematics, paying particular attention to how they inform our discussion on the limits of knowledge and the nature of reality. As we will see, the incompleteness of knowledge and the limits of our scientific worldview only add to the richness of our search for meaning, as they align science with our human fallibility and aspirations.

As I write these lines, an unknown choreography organizes the firing of millions of neurons in my brain; thoughts emerge and are expressed as words, typed on my laptop by a detailed coordination of eye and hand muscles. Something is in charge, an entity we loosely call "mind." I'm flying at about thirty thousand feet, returning from a documentary shooting in Los Angeles. The theme was the known universe, a retelling of the wonderful conquests of modern science, in particular those of astronomy and cosmology. I see the white clouds hovering below, the blue sky above; I hear the jet engines humming and feel the annoying tapping of my neighbor as he listens to music on his iPod.

My perception of the world around me, as cognitive neuroscience teaches us, is synthesized within different regions of my brain. What I call reality results from the integrated sum of countless stimuli collected through my five senses, brought from the outside into my head via my nervous system. Cognition, the awareness of being here now, is a fabrication of a vast set of chemicals flowing through myriad synaptic connections between my neurons. I am, and you are, a self-sustaining electrochemical network enacted across a web of biological cells. And yet we are much more. I am me and you are you and we are different, even if made of the same stuff. Modern science has removed the age-old Cartesian dualism of matter and soul in favor of a strict materialism: the theater of the self happens in the brain, and the brain is an assembly of interacting neurons firing nonstop like lights on a Christmas tree.

We have little understanding as to how exactly this neuronal choreography engenders us with a sense of being. We go on with our everyday activities convinced that we can separate ourselves from our surroundings and construct an objective view of reality. I know that I am not you, and I know that I am not the chair that I am sitting on. I can walk away from you and the chair, but I can't walk away from my own body. (Unless I'm in some trancelike state.) We also know that our perception of reality, on which we base our sense of self, is severely incomplete. Our senses capture only a sliver of what goes on around us. The brain is unaware of much that is happening, deaf and blind to a huge amount of information that wasn't particularly useful to increase our ancestors' survival chances in their hostile environments. For example, trillions of neutrinos racing all the way from the heart of the Sun zip through our bodies each second; electromagnetic waves of all sorts—microwaves, radio waves, ultraviolet, infrared—carry information we can't capture with our eyes; sounds beyond the range of our hearing go unheard; dust particles and bacteria go unseen. As the Fox said to the Little Prince in Antoine de Saint-Exupéry's fable, "What is essential is invisible to the eye."

Measuring instruments and tools greatly extend our view, whether of the very small or of the very far. They allow us to "see" invisible bacteria, electromagnetic radiation, subatomic particles, and exploding stars billions of light-years away. High-tech devices allow doctors to visualize tumors inside our lungs and brains, and geologists to locate underground oil reservoirs. Still, any detection or measuring technology has a limited precision or reach. A scale measures weight only with accuracy up to half its smallest graduation: if the ticks are spaced by one ounce, weights can be stated only to within a half-ounce precision. *There is no such thing as an exact measurement*. Every measurement must be stated within its precision and quoted together with "error bars" estimating the magnitude of errors. High-precision measurements are simply measurements with small error bars or high confidence levels; there are no perfect, zero-error measurements.

Consider now a less prosaic example than weighing scales: particle accelerators. These machines are designed to study the fundamental composition of matter, as they search for the smallest bits of stuff that make up all that exists in the world.[1] Particle accelerators make full use of Einstein's famous $E = mc^2$ formula, converting the energy of the motion of high-speed particles into new chunks of matter. As such, they work in a brutish way, colliding particles that move close to the speed of light head-on. How else could scientists study what exists, for instance, inside a proton? Unlike oranges, protons can't be cut. The solution is to throw protons against one another at high speeds and study the bits that come flying off after a collision. In the absence of a sharp knife, the internal composition of oranges could also be studied this way, tossing one fruit against another at high speeds and watching for the seeds, juice, and guts that come spewing out. Pushing the analogy a bit further: the speedier the oranges, the more telling the experiment. For example, only high-speed collisions would reveal the existence of seeds. A few collisions at higher speeds may even crack the seeds open, exposing their innards. This is an essential point: the higher the energy of the collision, the deeper we see into matter.[2]

During the past half century, particle accelerators underwent a tremendous escalation in power. The radioactive particles that Ernest Rutherford used in 1911 to probe the structure of the atomic nucleus had energies about a million times smaller than those currently achieved at the Large Hadron Collider, the behemoth particle accelerator located in Geneva, Switzerland. Consequently, modern-day particle physicists can probe deeply into the nature of matter, "seeing" things Rutherford wouldn't have dreamed of, such as "elementary" particles a hundred times heavier than a proton, like the famous Higgs boson, discovered in July 2012.[3] If funding is secured for future accelerators—a big if, given the enormous price tag for these machines—there is good reason to expect that new technologies will allow for the study of ever-higher energy processes and will produce exciting, perhaps even revolutionary, discoveries.

However, and this is a key point, technology limits how deeply experiments can probe into physical reality. That is to say, machines determine what we can measure and thus what scientists can learn about the Universe and ourselves. Being human inventions, machines depend on our creativity and available resources. When successful, they measure with ever-higher accuracy and on occasion may also reveal the unexpected. A case in point is Rutherford's surprise when his experiments revealed that even though the atomic nucleus occupies a tiny fraction of an atom's volume, it carries most of the atom's mass. To Rutherford and fellow scientists working during the early twentieth century, the world of atoms and subatomic particles looked very different from how it looks today. We can be equally sure that one hundred years from now our knowledge of subatomic physics will be quite different again. With my argument for now restricted to a purely empirical standpoint, scientists can grasp only what happens at energies within their experimental reach.

That being the case, what could we say *with certainty* about the properties of matter at energies thousands or millions of times higher than current limits? Theories may speculate about the properties of matter at such energies, and they may provide compelling arguments based on simplicity and elegance as to why things should be this way and not another. But the essence of empirical science is that Nature always has the last word: data cares little for our yearnings for aesthetic beauty, a point I explored in detail in my book *A Tear at the Edge of Creation*. It then follows that if we only have limited access to Nature through our tools and, more subtly, through our restricted methods of investigation, our knowledge of the natural world is necessarily limited.

Coupled to this technological limitation of how we probe the natural world, advances in physics, mathematics, and computation during the past two hundred years have taught us a lesson or two about the elusiveness of Nature. As we will see in detail, there are fundamental limits to how much we can know of the world not only because of our tools of exploration but also because Nature itself—at least as we humans perceive it—operates within certain limits. The Greek philosopher Heraclitus already sensed this twenty-five centuries ago, when he pronounced that "Nature loves to hide." Through countless tales of tribulation and success, we have found out that Nature cannot be beaten in this game of hide-and-seek. To paraphrase Samuel Johnson's frustration with trying to define certain English verbs, it is as if we were trying to paint a forest's reflection on the surface of a stormy lake.

As a consequence, and despite our ever-increasing efficiency, at any given time large portions of the natural world remain unseen or, more precisely, undetected. This shortsightedness, however, is a powerful tease to our imagination: limits should not be seen as insurmountable obstacles but as challenges. As the French prescient author Bernard le Bovier de Fontenelle wrote in 1686, "We want to know more than we can see."[4] Galileo's 1609 telescope could barely discern Saturn's rings, something that household telescopes can do today. What we know of the world is only what we can detect and measure. We see much more than Galileo, but we can't see it all. And this restriction is not limited to measurements: speculative theories and models that extrapolate into unknown realms of physical reality must also rely on current knowledge. When there is no data to guide intuition, scientists impose a "compatibility" criterion: any new theory attempting to extrapolate beyond tested ground should, in the proper limit, reproduce current knowledge. For instance, Einstein's theory of general relativity, which describes gravity as the curvature of spacetime as a result of the presence of matter (and energy), reduces to the old Newtonian theory of universal gravitation within the limit of weak gravitational fields: to launch spaceships to Jupiter we don't need Einstein's theory, but to describe black holes we do.

If large portions of the world remain unseen or inaccessible to us, we must consider the meaning of the word "reality" with great care. We must consider whether there is such a thing as an "ultimate reality" out there—the final substrate of all there is—and, if so, whether we can ever hope to grasp it in its totality. Note that I am refraining from calling this ultimate reality God, since God's nature, as most religions propose, is ungraspable. It is also not the subject of a scientific inquiry. And I am refraining from equating ultimate reality with any of the several Eastern philosophical notions of transcendent reality, as in a nirvana-like state achievable through meditation, the Brahman from Hindu Vedanta philosophy, or an all-encompassing Tao. For now, I am only considering the more concrete nature of *physical* reality, which we can infer through the diligent application of science. We thus must ask whether grasping reality's most fundamental nature is just a matter of pushing the limits of science or whether we are being quite naïve about what science can and can't do.

Here is another way of thinking about this: if someone perceives the world through her senses only (as most people do), and another amplifies her perception through the use of instrumentation, who can legitimately claim to have a truer sense of reality? One "sees" microscopic bacteria, faraway galaxies, and subatomic particles, while the other is completely blind to such entities. Clearly, they "see" different things and—if they take what they see literally—will conclude that the world, or at least the nature of physical reality, is very different. Who is right?

Asking who is right misses the point, although surely the person using tools can see further into the nature of things. Indeed, to see more clearly what makes up the world and, in the process, to make more sense of it and of ourselves is the main motivation to push the boundaries of knowledge, as de Fontenelle knew when he wrote, "All philosophy is based on two things only: curiosity and poor eyesight."[5] Much of what we do can be summed up as different attempts to alleviate our myopic gaze.

What we call "real" is contingent on how deeply we are able to probe reality. Even if there is such a thing as the true or ultimate nature of reality, all we have is what we can know of it. For the sake of argument, let us concede that sometime in the future a brilliant theory, supported by breakthrough experiments, will make remarkable inferences about the ultimate nature of reality. Even if we were to capture a glimpse of this reality through our detectors, all we would be able to conclude is that the theory makes partial sense: the tool-driven methodology we *must* use to learn about the world cannot prove or disprove theoretical statements about the ultimate nature of reality. To stress my point, our perception of what is real evolves with the instruments we use to probe Nature. Gradually, some of what was unknown becomes known. For this reason, what we call "reality" is always changing. The Earth-centered cosmos of Columbus was radically different from the Sun-centered Newtonian cosmos. The expanding cosmos of today, with billions of swirling galaxies, each with billions of stars, would have mystified Newton. It had even Einstein mystified. The version of reality we might call "true" at one time will not remain true at another.

Of course, Newton's laws of motion will always apply within their limit of validity, and water will always be composed of oxygen and hydrogen, at least within the particularly human way of describing physical and chemical processes with atoms. But these belong to our enduring explanations of the natural world, which are valid within their range of applicability and conceptual structure. Given that our instruments will always evolve, tomorrow's reality will necessarily include entities not known to exist today, whether astrophysical objects, elementary particles, or viruses. More to the point, as long as technology advances—and there is no reason to suppose that it will ever stop advancing for as long as we are around—we cannot foresee an end to this quest. The ultimate truth is elusive, a phantom.

Consider, then, the sum total of our accumulated knowledge as constituting an island, which I call the "Island of Knowledge." By "knowledge" I mean mostly scientific and technological knowledge, although the Island could also include all the cultural and artistic creations of humankind. A vast ocean surrounds the Island of Knowledge, the unexplored ocean of the unknown, hiding countless tantalizing mysteries. Whether this ocean extends toward infinity is something we will return to later on. For now, it is enough to imagine the Island of Knowledge growing as we discover more about the world and ourselves. This growth often takes an uncertain path, the coastline delineating the jagged boundary between the known and the unknown. Indeed, the growth may, on occasion, retrocede as ideas once accepted are jettisoned in light of new discoveries.

The Island's growth has a surprising but essential consequence. Naïvely, we would expect that the more we know of the world, the closer we would be to some sort of final destination, which some call a Theory of Everything and others the ultimate nature of reality. However, holding on to our metaphor, we see that as the Island of Knowledge grows, so do the shores of our ignorance—the boundary between the known and the unknown. Learning more about the world doesn't lead to a point closer to a final destination—whose existence is nothing but a hopeful assumption anyway—but to more questions and mysteries. The more we know, the more exposed we are to our ignorance, and the more we know to ask.[6]

Some people, including many of my scientist friends, consider this view of knowledge to be a downer. I've even been called a defeatist before, which is horrifyingly wrong, since I am proposing the precise opposite: a celebration of humanity's achievements resulting from our endless striving for new knowledge. "If we are never going to get to a final answer, what's the point of trying?" they ask. "And how do you know you are right, anyway?" This book will answer these questions. But just for starters, once we explore the nature of human knowledge—that is, how we try to make sense of the world and of our place in it—it should be obvious that our approach is fundamentally limited in scope. This realization should *open* doors, not close them, since it makes the search for knowledge an open-ended pursuit, an endless romance with the unknown. And what could be more inspiring than knowing that there will always be something new to discover in the natural world, that no matter how much we know there will always be plenty of room for the unexpected? To me, the real downer is to presume that there is an end to the search and that we will eventually get there. Paraphrasing Tom Stoppard in *Arcadia*, "It is the need to know that makes us matter."

New discoveries throw light here and there, but farther out their glow fades into darkness. How we choose to relate to this fact traditionally defines how most people approach life's mysteries: either reason will slowly but surely conquer the unknown, or it won't. If the latter, then something beyond reason is needed to cope with our perennial ignorance, such as belief in alternative explanations, which may include the supernatural. If these are the only two options we have, we are left with the unfortunate polarity between scientism versus supernaturalism that so much defines our age. I propose a third path, based on how an understanding of the way we probe reality can be a source of endless inspiration without the need for setting final goals or promises of eternal truths.

As science advances, we will know more. But we will also have more to know. New tools of exploration present us with new questions. Often, these are questions that couldn't even have been imagined before the tools were available. To take two obvious examples, consider astronomy prior to and after the telescope (1609) or biology prior to and after the microscope (1674): no one could have anticipated the revolutions these two instruments and their descendants would bring about. This unsettled existence is the very blood of science. Science needs to fail to move forward. Theories need to break down; their limits need to be exposed. As tools probe deeper into Nature, they expose the cracks of old theories and allow new ones to emerge. However, we should not be fooled into believing that this process has an end. The scientific approach to knowledge has essential limitations; some questions are beyond its reach. In fact, some key aspects of Nature will necessarily remain unknown to us. Some, I will argue, are unknowable.

To expose the limits of science is far from being an obscurantist; quite the contrary, it is a much needed self-analysis in a time when scientific speculation and arrogance are rampant. By describing the limits of explanations of physical reality based on scientific methodology I am attempting to protect science from attacks on its intellectual integrity. I am also trying to explain how science advances because of our ignorance and not because of our knowledge. As Columbia University neuroscientist Stuart Firestein remarked in his recent book *Ignorance: How It Drives Science*, grant proposals are first and foremost statements of the current state of ignorance. To claim to know the "truth" is too heavy a burden for scientists to carry. We learn from what we can measure and should be humbled by how much we can't. It's what we don't know that matters.

Our perception of reality relies on an artificial separation between subject and object, a question that has inspired (and confused) thinkers of all ages. You may think you know where you end and the "outside" world begins, but the issue, as we shall see, is far from simple. No two people have the exact same perspective on the world. On the other hand, science is the best toolkit we have to create a universal language that transcends individual differences. As we explore our longing to conquer the unknown, we will also experience science's power to transform and to inspire.

PART I

THE ORIGIN OF THE WORLD AND THE NATURE OF THE HEAVENS

In the beginning, God created the earth,
    and He looked upon it in His cosmic loneliness.
And God said, "Let Us make living creatures out of mud,
    so the mud can see what We have done."
And God created every living creature that now moveth,
    and one was man. Mud as man alone could speak.
God leaned close to mud as man sat, looked around,
    and spoke. "What is the purpose of all this?"
    he asked politely.
"Everything must have a purpose?" asked God.
"Certainly," said man.
"Then I leave it to you to think of one for all this,"
    said God.
And He went away.

    —KURT VONNEGUT, *CAT'S CRADLE*

It is questions with no answers that set the limits of
human possibilities, describe the boundaries of human
existence.

    —MILAN KUNDERA,
    THE UNBEARABLE LIGHTNESS OF BEING

Man has always been his own most vexing problem.
    —REINHOLD NIEBUHR,
    THE NATURE AND DESTINY OF MAN

CHAPTER 1

THE WILL TO BELIEVE

*(Wherein we explore the role of belief and extrapolation
in religion and in scientific creativity)*

C an we make sense of the world without belief? This is a central
question behind the science and faith dichotomy, one that in-
forms how an individual chooses to relate to the world. Contrasting
mythic and scientific explanations of reality, we could say that religious myths attempt to explain the unknown with the unknowable
while science attempts to explain the unknown with the knowable.
Much of the tension stems from assuming that there are two mutually inconsistent realities, one within this world (and thus "knowable" through the diligent application of the scientific method) and
one without (and thus "unknowable" or intangible, traditionally
related to religious belief).[1]

In myths, the unknowable reflects the sacred nature of the gods,
whose existence transcends the boundaries of space and time. In
the words of historian of religion Mircea Eliade:

For the Australian as well as for the Chinese, the Hindu and the
European peasant, the myths are *true* because they are *sacred*, because they tell him about sacred beings and events. Consequently,
in reciting or listening to a myth, one resumes contact with the

sacred and with reality, and in so doing one transcends the profane condition, the "historical situation."[2]

Throughout the ages, religious myths have allowed the faithful to transcend the "profane condition," the perplexing awareness we humans have of being creatures bound by time, of having a history and an end. At a more pragmatic level, mythic explanations of natural phenomena were prescientific attempts to make sense of things that were beyond human control, answering questions that seemed unanswerable. Why should the Sun go across the sky every day? To the Greeks, because Apollo transported it in his fiery chariot. To the Navajos of the American Southwest, it was Jóhonaa'éí who hauled the Sun daily on his back across the sky. To the Egyptians, this task belonged to Ra, who transported the Sun in his boat. In a strictly naturalistic sense, the motivation behind such myths is not so different from that of science, as both attempt to uncover the hidden mechanisms behind natural phenomena: after all, gods and physical forces make things happen, albeit in very distinct ways.

More to the point, both the scientist and the faithful *believe* in unexplained causation, that is, in things happening for unknown reasons, even if the nature of the cause is completely different for each. In the sciences, this belief is most obvious when there is an attempt to extrapolate a theory or model beyond its tested limits, as in "gravity works the same way across the entire Universe," or "the theory of evolution by natural selection applies to all forms of life, including extraterrestrial ones." These extrapolations are crucial to advance knowledge into unexplored territory. The scientist feels justified in doing so, given the accumulated power of her theories to explain so much of the world. We can even say, with slight impropriety, that her faith is empirically validated.[3]

Here is an example. Newton's theory of universal gravitation, as explained in Book III of his revolutionary *Mathematical Principles of Natural Philosophy*, the *Principia*, should really have been called a theory of solar system gravitation, since by the late seventeenth century no tests were conceivable beyond its confines. Yet Newton called Book III *The System of the World*, assuming that his description of gravitational attraction as a force proportional to the quantity of mass in two bodies and decreasing with the square of the distance between them would extend to the whole "world," that is, the cosmos. In his own words, from Book III,

> Finally, if it is universally established by experiments and astronomical observations that all bodies on or near the earth gravitate toward the earth, and do so in the proportion of matter in each body, and that the moon gravitates toward the earth in proportion to the quantity of its matter, and that our sea in turn gravitates toward the moon, and that all planets gravitate toward one another, and that there is a similar gravity of comets toward the sun, it will have to be concluded by this third rule that all bodies gravitate toward one another.[4]

Newton cleverly avoided speculating on the cause of gravity itself—"I feign no hypothesis"—attaching it universally to all bodies with mass: "And to us it is enough that gravity does really exist, and acts according to the laws which we have explained, and abundantly serves to account for all the motions of the celestial bodies, and of our sea," he wrote in the General Scholium of the *Principia*, a sort of concluding explanatory text. He didn't know why masses attract one another, but he knew how they did so. The *Principia* was a book concerned with the hows and not with the whys.

Later, in a letter to the Cambridge theologian Richard Bentley dated December 10, 1692, Newton used his extrapolation on the nature of the gravitational force to justify why the universe should be infinite, a major turning point in the history of cosmological thought. If gravity acted across a spatially finite universe according to the same law of attraction, Bentley wondered, why wouldn't all matter be concentrated in a huge ball at the center? Newton agreed that this would indeed be the case if the universe were finite in extent. However, he went on, "if the matter was evenly diffused through an infinite space, it would never convene into one mass but some of it convene into one mass and some into another so as to make an infinite number of great masses scattered at great distances from one to another throughout all that infinite space." Newton's belief in the universal nature of gravity was strong enough to let him speculate confidently about the spatial extent of the cosmos as a whole.

Centuries later, Einstein did something similar. He formulated his general theory of relativity in final form in 1915, wherein he went a step beyond Newton and attributed gravity to the curvature of space about a massive body (and time, but let's leave this aside for now): the larger the mass, the more space is bent around it, like the elastic surface of a trampoline around people of different weights. No more was a mysterious action-at-a-distance called forth to explain how massive bodies tend toward one another: in a curved space trajectories are no longer straight. Of course, Einstein didn't explain why mass should have this effect on the geometry of space. I suspect that, like Newton, he would have answered that he "feigned no hypothesis." His theory worked beautifully, explaining things Newton's couldn't, as observational tests concerned with solar system dynamics attested. And that was enough.

In 1917, less than two years after the publication of his general theory, Einstein wrote a remarkable paper, "Cosmological Considerations on the General Theory of Relativity." Like Newton, Einstein extrapolated the validity of his theory beyond the solar system, where it was tested at the time, to the universe as whole, and he proceeded to consider the shape of the entire cosmos. In true Platonist fashion, he wanted the cosmos to have the most perfect of shapes, that of a sphere. For convenience, and because of the lack of any opposing observation at the time, he also wanted the universe to be static. His equations produced the desired answer—a static and spherically symmetric Universe—but with a hidden surprise: in order to avoid the total collapse of matter to a central point (which would occur, just as Bentley had worried in Newton's case), Einstein refrained from making space infinite in extent. Instead, he introduced what he called a "universal constant" and added a new term to the equations describing the curvature of space, noting that, if sufficiently small, this constant was "also compatible with the facts of experience derived from the solar system." This constant, "not justified by our actual knowledge of gravitation," he conceded, is now called the "cosmological constant" and may indeed play a key role in the dynamics of the cosmos, although one quite different from that which Einstein had prescribed. Einstein needed it to ensure that his static spherical universe would not collapse onto itself. Displaying complete faith in his theory, he not only extrapolated his equations from the solar system to the entire universe but also imposed on his theory of the cosmos the effects of a strange repulsion whose job was to balance the cosmic dome.

To go beyond the known, both Newton and Einstein had to take intellectual risks, making assumptions based on intuition and personal prejudice. That they did so, knowing that their speculative theories were necessarily faulty and limited, illustrates the power of belief in the creative process of two of the greatest scientists of all time. To a greater or lesser extent, every person engaged in the advancement of knowledge does the same.

CHAPTER 2

✕✕✕

BEYOND SPACE AND TIME

(Wherein we explore how different religions have faced
the question of the origin of all things)

Backtrack ten thousand years, to just before the dawn of the first great civilizations along the Tigris and the Euphrates Rivers, where Iraq is now. To divinize Nature was an attempt to have a certain measure of control over what was uncontrollable. Floods, droughts, earthquakes, volcanoes, tidal waves—what even today insurance companies (shamelessly) call "acts of God"—were attributed to angry gods who needed to be placated. A language had to be developed, a common dialect between human and deity, enacted through ritualistic practices and mythic narratives, to bridge the enormous power imbalance between humans and the forces of Nature. As threats to survival came from everywhere—within the Earth, from its surface, and from the skies—gods had to be everywhere as well. Religion was born of necessity and reverence. Quite possibly, any thinking being with widespread but limited powers must assume the existence of other beings, be they gods or, more recently, aliens, with powers beyond his. The alternative, to leave natural disasters to chance, was just too scary to contemplate, as it would imply in accepting humankind's helplessness and utter loneliness in confronting the unknown. To have a fighting chance to control their destiny, humans had to believe.

Fear was not the only driving force toward belief, although it was possibly the main one. But not everything was bad. Good things also happened: a good crop, a productive hunt, quiet weather, bountiful oceans. Nature didn't only take away; it also gave plenty. In its dual role as giver and taker, it kept people alive, and it could kill them. Reflecting this polar tension, natural phenomena could be regular and safe—the day-night cycle, the seasons, the phases of the Moon, the tides—or irregular and fearsome, as in solar eclipses, comets, avalanches, and forest fires. It is then not surprising that regularity was (and is) associated with good and irregularity with bad: natural phenomena gained a moral dimension that, through the divinization of Nature, reflected directly the whims of intangible gods.

Across the world, ancient cultures erected monuments and temples to celebrate and to clock the regularity of the heavens. In England, Stonehenge's function as a burial place is probably related to the yearly alignment of the "Heel Stone" with the rising Sun during summer solstice, establishing a link between the periodic return of the Sun and the human's cycle of life and death. If the mechanisms behind the cyclic motions in the heavens were unknown—and there was no desire to "know" them, at least in the way we understand knowing now—they were still noted and in some cases measured with great care. Some three thousand years ago, the Babylonians, for example, had a well-established astronomical tradition, reflected in their creation myth, the Enuma Elish ("When Above"). They made detailed tables mapping the motions of the planets and the Moon across the sky, and registered any observed periodicities, as in the Ammisaduqa tablet, which recorded the risings and settings of Venus for twenty-one years.

There is comfort in repetition. If Nature beats to a drum, perhaps we do too. A cyclic time brings with it the promise of rebirth, establishing a deep connection between human and cosmos: our existence reflects that of the whole world. No wonder the myth of the eternal return resonates with so many cultures. What could be better than to believe that we return over and over again, that death is not the end but a transition to a new beginning?

As a father of five, I see this struggle with endings in all my children. My son Lucian, who was six when I wrote these lines, has been obsessed with death since he was four. Death sounds like an absurdity when time seems endless. "What happens after we die?" is one of those questions every parent hears, and most struggle to answer. Lucian is convinced that we return. He is just not quite sure if we return the same or as a different person. His choice, of course, is to come back the same, with the same parents and siblings, essentially reliving life twice or, better still, endlessly. What could be safer than not to have to face loss? It breaks my heart to have to tell him that what happens to us is the same thing that happens to the ant he crushes under his feet. He, of course, is not convinced. "How do you know, Dad?" "I don't know for sure, son. Some people believe we do come back; others that we go to a place called Paradise, where we meet everybody else who has died. The problem is that I haven't heard back from any of them to be sure that that's where we are headed." The conversation usually ends with a very tight hug and many utterings of "I love you." What could be harder than to know that I cannot love him forever? And that one day, in the normal course of things, he will have to cope with my death?

With the advent of the Abrahamic faiths, a radically different way to think about the nature of time made a triumphal entrance: instead of ongoing cycles of creation and destruction, of life and death, time becomes linear, with a single beginning and an end. "Profane history," as Eliade called it, is what happens between birth and death. The stakes suddenly became much higher, since with a single lifetime there is only one chance to be happy. For Christians and Muslims, the notion of an after-death Paradise comes to the rescue, and time begets a dual role, linear in life and inexistent in Paradise.

Linear or cyclic, time has always been a measure of transformation. Follow it to the future; it leads to endings. Follow it to the past; it leads to beginnings. In mythic narratives, humans are always subjected to the changes that time brings, while gods live outside time, never aging or getting sick. As life begets life, and generations succeed one another, following time backwards will necessarily lead to first life, the first living thing, be it bacteria, human, or beast. It is here the key question arises: How did the first living creature emerge, if there was nothing already living to give it birth? The same reasoning can be extrapolated to the world: How did the world come to be, if it had a beginning? The mythic answer, in the vast majority of cases, is clear: gods created the world first and then life. Only that which exists without time can *first* create that which exists within time. Although some creation myths, most notably among the Maori of New Zealand, suggest that the first creation could have happened without the interference of gods, in most myths time itself becomes a creation, starting once the world comes into being, as Saint Augustine cleverly proposed in *The Confessions* (Book 11, Chapter 13):

> Seeing then Thou art the Creator of all times, if any time was before Thou madest heaven and earth, why say they that Thou didst forego working? For that very time didst Thou make, nor could times pass by, before Thou madest those times. But if before heaven and earth there was no time, why is it demanded, what Thou then didst? For there was no "then," when there was no time.

The origin of the world and the beginning of time are thus deeply enmeshed with the nature of the heavens, a connection that remains true in our time as modern cosmological models attempt to describe the origin of the Universe and astrophysicists study the origin of stars and planets. Not surprisingly, as I have examined in my book *The Dancing Universe*, both cyclic and linear notions of time reappeared in modern cosmology. More surprisingly, an essential characteristic of ancient creation myths—the deep relation between human and cosmos—also returns with current astronomical thought, after a long post-Copernican hiatus when our existence played second fiddle to the material splendor of the Universe. When Copernicus and, more pointedly, Johannes Kepler and Galileo Galilei displaced Earth from the epicenter of Creation during the first decades of the seventeenth century, we lost our special status to become mere inhabitants of one among countless many worlds. Four hundred years later, as the ongoing search for life in the Universe reveals the fragility and relative scarcity of Earthlike planets, life and, more crucially, the uniqueness of human life is regaining its cosmic relevance: we matter because we are rare. The many steps from nonlife to life and then to complex multicellular life are hard to duplicate. Furthermore, the many particulars depend on our planet's detailed history. However, even with the current lack of evidence we cannot establish conclusively that other kinds of intelligent life don't exist in the Universe. They may or may not be out there. But what we *can* do is to state with confidence that *if* intelligent aliens exist, they are distant and rare. (Or, if ubiquitous, they certainly know how to hide extremely well, something we will get to at the end of this book.) In effect we are alone and must learn to live with our cosmic loneliness.

The urge to know our origins and our place in the cosmos is a defining part of our humanity. Creation myths of all ages ask questions not so different from those scientists ask today, when they ponder the quantum creation of the Universe "out of nothing," or whether our Universe is but one among countless others, all of them exhalations of a timeless multiverse. The specifics of the questions and of the answers are, of course, entirely different, but not the motivation: to understand where we came from and what our cosmic role is, if any. To the authors of those myths, ultimate questions of origins were solely answerable through invocations of the sacred, as only the timeless could create that which exists within time. To those who do not believe that answers to such questions remain exclusively within the realm of the sacred, the challenge is to scrutinize the reach of our rational explanations of the world and examine how far they can go in making sense of reality and, by extension, of ultimate questions of origins.

CHAPTER 3

TO BE, OR TO BECOME?
THAT IS THE QUESTION

*(Wherein we encounter the first philosophers of ancient Greece
and delve into their remarkable notions about
the meaning of reality)*

A major shift in perspective happened sometime during the sixth and fifth centuries BCE in ancient Greece. Although influential new ideas concerning humankind's social and spiritual dimension were appearing elsewhere, such as with Confucius and Lao Tzu in China, and with Siddhartha Gautama, the Buddha, in India, it is to Greece that we turn to witness the birth of Western philosophy, a new mode of understanding based on questioning and argumentation, devised to examine the fundamental nature of knowledge and existence. In opposition to creation myths and sacred knowledge based on intangible revelation, the first Greek philosophers, known collectively as the Presocratics (since most lived before Socrates), sought to understand reality through logic and conjecture. This transition, in which tremendous faith is placed in the powers of reasoning to address the key questions of existence, redefined humanity's relation with the unknown from a passive reliance on fate and the supernatural to an active approach to knowledge and personal freedom.

Central to the Ionians, the first group of Presocratic thinkers, was a preoccupation with the material composition of the world. "What is the stuff that makes everything that is?" they asked. That this remains the defining question of modern particle physics serves to show that the value of a great question is that it keeps generating answers that, in turn, keep changing as our methods of inquiry change. Different members of the Ionian school suggested different answers, although all shared a fundamental characteristic: they believed that "All is One," that is, the material essence of reality is captured in a single substance or entity. This centralizing Oneness stands in sharp contrast with previous pantheistic mythologies, in which different gods are responsible for different parts of Nature. To the Ionians, everything that we witness is a manifestation of a single material essence undergoing various types of physical transformation.

And so Thales, whom none other than Aristotle considered to be the first philosopher, is said to have declared that "the principle of all things [is] water. For [Thales] says from water come all things and into water do all things decompose."[1] This quote, from the Byzantine physician Aëtius of Amida, is typical of thoughts attributed to Thales. Unfortunately, none of Thales's works survived, and we must rely on indirect sources to make sense of his ideas. When we read the literature, it becomes clear that Thales did propose water to be the source of everything, recognizing its role as the giver of life. To him, water symbolized the ongoing transformations of Nature, never at rest, even when apparently so. To explain the power source for these transformations, Thales invoked a soul-like force: "Some say that [soul] is mixed in the totality; this is perhaps the reason Thales thought all things are full of gods," wrote Aristotle in *On the Soul*.[2] These, however, are not the anthropomorphic gods of past mythologies, but the unexplained powers behind the changes witnessed in physical reality. Thales, and all of the Ionians, preached a philosophy of becoming, of ongoing transformation springing from a single material source: everything comes from it, and to it everything returns.

It is quite remarkable that even if the first philosophers of the West lived in a culture with a belief system that relied on a multiplicity of gods to do different things, they searched for a single explanation for reality, an absolute principle for existence. They were, quite explicitly, looking for a unified theory of Nature, the first Theory of Everything. Much later, the historian of ideas Isaiah Berlin called this belief in Oneness, which survives to this day, the "Ionian Fallacy," declaring it to be meaningless: "A sentence of the form 'Everything consists of . . . ' or 'Everything is . . . ' or 'Nothing is . . . ' unless it is empirical . . . states nothing, since a proposition which cannot be significantly denied or doubted can offer us no information."[3] In other words, authoritative all-encompassing statements, which cannot be comparatively contrasted or measured, carry no information: they are articles of faith, not reason. I will have much to say about this later on, when we examine the search for ultimate explanations in science. For now, we must investigate the protoscientific thinking of Thales's immediate successor, Anaximander of Miletus, rightfully considered the first scientific philosopher, for his conceptualization of Nature in mechanistic terms.

Instead of a concrete material substance Anaximander went abstract, proposing that some primordial medium, "the boundless" (*apeiron*) was the source of all things: "For from this all things come to be and into this all things perish. That is why countless world-orders are generated and again perish into that from which they came to be," wrote Aëtius, summarizing Anaximander's thoughts.[4] The boundless is thus uncreated and indestructible, the primal material principle, which has existed for all eternity and through the infinity of space.

Anaximander saw the world as a chain of events following from natural causes.[5] According to many sources, in his lost treatise *On*

Nature, the first known text in natural philosophy, Anaximander attempted to explain all sorts of phenomena, from lightning (which comes from the movement of air in clouds) to the origin of humans (from life that first originated at sea and then migrated to land). In the words of Daniel W. Graham, "Whatever he may have learned from Thales, Anaximander was a true revolutionary who, by organizing insights into a cosmogonical theory and contributing them to papyrus, presented a framework for thinking about Nature as an autonomous realm with elemental entities and laws of interaction. As far as we can tell, he was the founder of scientific philosophy."[6]

Instead of Apollo pulling the Sun across the sky in his chariot, Anaximander proposed that the Sun was a hole in a fiery wheel that revolved around the Earth. Simplistic as this idea may sound to us, its historical importance cannot be exaggerated: this is the first mechanical model of the heavens, an attempt to explain the perceived motions of celestial objects through a pattern of cause and effect without divine interference. Anaximander's depiction of the origin of the cosmos was equally mechanistic and imaginative. As Plutarch reported in his *Miscellanies*, "[Anaximander] says that that part of the [boundless] which is generative of hot and cold separated off at the coming to be of the world-order and from this a sort of sphere of flame grew around the air like bark around a tree. This subsequently broke off and was closed into individual circles to form the Sun, the Moon, and the stars."[7] That is, not just the Sun but also the Moon and stars were punctures in fiery interlocked wheels revolving around the Earth. The cosmos became an ordered mechanism, following the rules of cause and effect.

Anaximander's ideas, as well as all ideas coming from the Presocratics and subsequent Greek philosophers, were simply based on intuition and the power of argumentation, and thus lacked any connection with the notion of testability. Nevertheless, they stand out for their remarkable intellectual courage and imagination. The Greeks were not the first to ask questions about cosmic origins or the nature of reality. But differently from their predecessors, they set a new agenda for humanity, wherein the freedom to ponder such questions was a fundamental right of any individual and the only path to intellectual independence and personal happiness.[8] As Lucretius famously wrote in *On The Nature of Things* (50 BCE), his long narrative poem based on the philosophy of the Presocratic Atomists Leucippus and Democritus, and one of the most lucid defenses of atheism ever composed:

> This terror, then, this darkness of the mind,
> Not sunrise with its flaring spokes of light,
> Nor glittering arrows of morning can disperse,
> But only Nature's aspect and her law,
> Which, teaching us, hath this exordium:
> Nothing from nothing ever yet was born.
> Fear holds dominion over mortality
> Only because, seeing in land and sky
> So much the cause whereof no wise they know,
> Men think Divinities are working there.
> Meantime, when once we know from nothing still
> Nothing can be created, we shall divine
> More clearly what we seek: those elements
> From which alone all things created are,
> And how accomplished by no tool of Gods.[9]

The sharp separation Lucretius advocates between a rational approach to understanding the world and a belief in active deities was not widespread. Quite the contrary, for many of the Presocratics, and certainly for Plato and Aristotle later, we see a link between the two: the cosmos only made sense with a coexisting godlike presence. In no Presocratic school is this more transparent than with the Pythagoreans, the caste of number mystics who believed that the essence of Nature was cyphered in combinations (ratios) of integers like 1, 2, 3, and so on. Settled in southern Italy, and thus geographically far from the Ionians of Western Turkey, the Pythagoreans believed that the path toward enlightenment was forged through an understanding of mathematics and geometry, the tools the architect deity used to construct the cosmos.

According to the writings of Philolaus of Croton, a prominent follower of Pythagoras who lived around 450 BCE, the center of the cosmos was not Earth or the Sun but the "central fire," the Citadel of Zeus. This pre-Copernican displacement of Earth's centrality was justified through theological arguments: only God could occupy the center of all things. As Aristotle reported in his *On the Heavens*, "Most people say that the Earth lies at the center of the universe . . . but the Italian philosophers known as the Pythagoreans take the contrary view. At the center, they say, is fire, and the Earth is one of the stars, creating night and day by its circular motion about the center."[10] This remarkable notion may well have influenced later thinkers who proposed to displace the Earth from the center of the cosmos, most notably Aristarchus of Samos around 280 BCE and, famously in the sixteenth century, Nicolas Copernicus. In Copernicus's own words, from his *On the Revolutions of the Heavenly Spheres*:

> And in fact first I found in Cicero that Hicetas supposed the Earth to move. Later I also discovered in Plutarch that certain others were of this opinion. I have decided to set his words down here, so that they may be available to everybody: *some think that the Earth remains at rest; but Philolaus the Pythagorean believes that, like the Sun and Moon, it revolves around the fire in an oblique circle.* . . . Therefore, having obtained the opportunity from these sources, I too began to consider the mobility of the Earth.[11]

The roots of the so-called Copernican revolution extend much farther into the past than is generally believed.

Most of us encountered the Pythagorean theorem in high school, the one relating the three sides of a right triangle. Even though the legendary sage Pythagoras is credited with its discovery, it may have been a credit by authority: someone else discovered it, but the master got the credit. In any case, Pythagoras apparently did discover what might be considered the first mathematical law of Nature, a relation between the sounds of a musical scale and the lengths of the strings that produce them. He realized that the sounds that feel harmonious to the ears correspond to simple ratios of string lengths, ratios that contain only the first four integer numbers 1, 2, 3, and 4, which constituted their sacred *tetractys*, "the fount and root of ever-flowing Nature," as later wrote Sextus Empiricus, referring to the key idea of the Pythagoreans. For example, if the length of a string is L, plucking it at half length (L/2) gives a sound that is an octave higher than the string at full length; at two-thirds length (2L/3) gives a fifth; at (3L/4), a fourth. Since what feels harmonious is a gateway to the inner workings of the psyche, Pythagoras and his followers found a bridge between the outside world and its perception through the senses. That this bridge was built with mathematical relations established the foundation to what would come next: to understand the world we must describe it mathematically. More than that, since what is harmonious is beautiful, the beauty of the world is expressible through mathematics. A new aesthetics emerges, equating mathematical law with beauty and beauty with truth.

Apart from pioneering the role of mathematics in the description of the world and of our interactions with it, the Pythagoreans had plenty to contribute to cosmology. The Pythagorean cosmos did more than displace the Earth from the center of the cosmos in favor of the "central fire." Extrapolating from harmony in music to the celestial spheres, the Pythagoreans believed that the distances between the planets stood in the same numerical ratios as those of the musical scales. As the planets revolved in the heavens they made music, "the harmony of the spheres," although according to legend only Pythagoras himself was able to hear it. The cosmic design, from the sensorial pleasure of music to the distances between planets, embodied harmoniously strict proportions: the beauty of creation was in its essence mathematical. Nothing could be more ennobling to the human soul than to decipher it.

Before we transition into Plato and his pupil Aristotle, we should briefly review where we are. On the one hand, we have the Ionians proposing that the essence of Nature is transformation and that all that exists is a manifestation of a single material essence. On the other, we have the Pythagoreans proposing that mathematics offers the key to Nature's mysteries and to our perception of reality. There were other voices. Also from Italy, Parmenides and the Eleatics countered the Ionians, proposing that what is essential is not that which can change but that which is unchangeable, or "what is." They proposed that change is an illusion, a distortion our imperfect senses cause in how we perceive reality. These are some of the earliest considerations in the West about the nature of reality and our perception of it: Is the essence of reality "in your face," the transformations we witness with our senses? Or is its essence hidden, locked in an abstract realm perceived only through thought?

To perceive change we need to sense it. But if our senses feed us only imperfect reconstructions of what exists, how can we grasp what is truly real? On the other hand, if we follow Parmenides, how can we possibly have any idea of this "thing" that doesn't change? After all, if something doesn't change, it becomes imperceptible to us, like a humming to which we grow deaf. Worse, if this unchangeable reality exists somehow in a rarefied realm, how are we to make sense of it? How can we probe it? And so the Ionians would accuse the Eleatics of empty abstractions, while the Eleatics would think the Ionians fools, as they trusted what could not be trusted. Meanwhile, the Pythagoreans would ignore both, following their belief in the power of mathematics to describe the harmony and beauty of the world.

Taken together, the plurality of Presocratic worldviews is staggering. The first Western philosophers were pushing the boundaries of knowledge in all directions, increasing the territory on which rational debate could ensue. Already twenty-five centuries ago conflicting ideas about the nature of reality abounded and clashed with one another. In their richness and complexity, they carried the kernel of the question that is still with us today and is the central theme of this book: To what extent can we make sense of reality? The Island was growing fast, and the shores facing the unknown offered endless possibilities.

CHAPTER 4

LESSONS FROM PLATO'S DREAM

(Wherein we explore how Plato and Aristotle
dealt with the question of the First Cause and with
the limits of knowledge)

B oth Parmenides and the Pythagoreans deeply influenced Plato, who lived between circa 428 and 348 BCE. In a sense, Plato unified their modes of thinking, since, like Parmenides, he despised sensorial experience as a reliable source to attain the truth while, like Pythagoras, he embraced geometrical notions as the bridge between the human mind and the world of pure thought, where this elusive truth was to be found. Living in a time of political and social unrest, when Athens finally fell to Sparta at the end of the Peloponnesian War in 404 BCE, Plato searched for unchangeable truths as a path to stability and wisdom.

Nowhere is Plato's mode of thinking as clear as in his famous Allegory of the Cave, which appears in *The Republic*. The Allegory is also one of the first explicit meditations on the nature of reality. Imagine a group of people inside a cave, chained and unable to move since birth. Let's call them the Chained Ones. All the Chained Ones could do was to stare at the cave wall in front of them. They had no knowledge of the world outside or around them; their "reality" consisted of what they saw projected on the cave wall. They were unaware of the large fire burning behind them and of the

25

small path and low-lying wall standing between them and the fire. Other persons could walk along the path and hold statues and various other objects near the fire. The Chained Ones saw the shadows of the statues and objects projected on the cave wall and took them for real. Their inability to turn back and understand their condition precluded them from seeing the truth. Theirs was a world of false illusions.

Plato proposed that even if a Chained One were freed and able to move toward the burning fire and the moving statues, the pain and temporary blindness from the bright flames would be such that he would quickly return to his previous spot in the cave. He would choose to believe the shadows to be truer than the blinding truth that was revealed to him: knowledge comes with a price that not all are willing to pay. To learn takes courage and tolerance, as it may cause an uncomfortable change of perspective. Plato insisted that if the freed Chained One had been dragged out of the cave and into the sunlight and hence even closer to the truth, he would beg to return to the shadows on the cave wall.

Plato compared the ascension of the Chained One toward the sunlight with the "mounting of the soul to the intellectual region," that is, with a literal enlightenment. He further suggested that the truth—a derivative of what he called "the essential Form of Good"—is hard to grasp, given that we are chained to our sensorial perception of reality. However, once we are ready to see (what we can of) it, the drive toward more knowledge is inevitable:

My opinion is that in the world of knowledge the idea of good appears last of all, and is seen only with an effort; and, when seen, is also inferred to be the universal author of all things beautiful and right, parent of light and of the lord of light in this visible world, and the immediate source of reason and truth in the intellectual; and that is the power upon which he who acts rationally either in public or private life must have his eyes fixed.[1]

Plato's main task in *The Republic* was to propose how a just and equitable society should be ruled and what kind of man should rule it. His answer, the philosopher-king, would ideally be someone who could probe into the abstract realm of pure forms, someone who would "set this Form of the Good before his eyes" and feed from the undying light of wisdom that forever shines there.

Plato's Forms are a source of much debate and confusion. Fortunately, we need not bother with the details. Suffice it to think of Forms as some kind of blueprint of perfection, as the core ideas behind things or feelings. For example, the Form of chairness contains in it the possibility of all chairs, which, once built, become mere shadows of the true Form. Forms are the universal essence of what is potentially existing, as they themselves are nonexistent in space and time. As limited beings, we can only dimly perceive what they comprise of, as we clumsily attempt to represent them within our perceived reality. Thus the idea of a circle, the one we sustain in our minds when prompted to think of a circle, is the only real circle: drawings or other forms of concrete representation are never the real thing, as they are never perfect.

In *Timaeus*, Plato extended these notions to cosmology. The Universe is the handiwork of a godlike entity called the demiurge, who uses Forms as a blueprint for his creation: the cosmos is spherical, and motions are circular and uniform, as those are the "most appropriate to mind and intelligence." Plato was proposing a cosmic aesthetics, the most symmetric and thus perfect shape being the only possible choice for celestial objects and their motions. Mind dictates the paths that matter follows: the world comes from an idea, and its physical structure must reflect this idea. This is a teleological cosmic view, a cosmoteleology, the notion that either the Universe has a purpose of its own or that it reflects the purpose of its Creator. It clashes frontally with the Atomistic notion of cosmic purposelessness, which, instead, proposes that nothing happens as the result of a preprescribed reason or intention: everything comes from atoms randomly moving and combining in the Void. As Lucretius wrote in *The Nature of Things*:

> Especially since this world is the product of Nature,
>   the happenstance
> Of the seeds of things colliding into each other by pure chance
> In every possible way, no aim in view, at random, blind,
> Till sooner or later certain atoms suddenly combined
> So that they lay the warp to weave the cloth of mighty things:
> Of earth, of sea, of sky, of all species of living beings.[2]

Most of the philosophical discussions on the nature of the Universe after Plato, all the way to present-day ones involving the possibility of a multiverse that encompasses countless many universes including our own, or the possibility that our existence has some kind of cosmic purpose, reflect this charged ancient dichotomy.

From a scientific perspective, the main obstacle to any teleological explanation is our inability to determine whether it is right or wrong. The scientific method relies on empirical validation: a scientific hypothesis must be falsifiable; that is, scientists must be able to prove it wrong. If we can't, or better, while we can't—since every hypothesis must fail sooner or later—we accept it as true.[3] So if someone states that "the Universe has a purpose," we must first identify this purpose and then verify if indeed it is there. A popular contender is conscious life: "the Universe has the explicit intent of creating conscious life." A Creator Universe is not so different from a Creator God. It transforms a supernatural teleology into a teleology of the supranatural. A Universe with a scientifically justified intent is the modern answer to the pressure that the countless successes of science have imposed on revelation-based explanations: it gives purpose a scientific credibility. A Universe that intentionally engenders conscious beings reflects our ever-present need to be not just special creatures but special creations.

Attractive as it may be, the question of purpose has a serious challenge. How are we to test purpose in Nature? Unless we receive a very clear message from the perpetrator(s) stating their purpose and thus resolving the issue, this kind of naturalistic teleology becomes a categorical unknowable: if there is cosmic purpose and we cannot be aware of it, we cannot know whether it is not there either. We are ignorant either way, and all we can do is choose to believe in it or not, just as Plato believed in his demiurge but could not prove his existence.

When Plato's pupil Aristotle came into the game, his goal was diametrically opposite to that of his master. Aristotle was in many ways a pragmatist; he tried to build a towerlike structure of interlinked rational arguments to explain the way Nature works. This verticalism would, of course, be extremely attractive to the church later on, as it adopted Aristotle's cosmic arrangement as its own. Aristotle would posit that the natural bottom-up vertical arrangement of the four basic substances—earth, water, air, and fire—explains why an object made of one of them, when displaced from its medium, would naturally move back to it: an air bubble will float upward in water until it mixes with air, while a rock will sink to the bottom of a lake. He dismissed Plato's Forms and artisan demiurge as mere abstractions, suggesting instead that teleological drive was to be found within the objects themselves, in their "natures," which he took to be principles of change inherent in living things.

However, Aristotle wouldn't dismiss godlike presences in the world. Even though he would posit that the Universe was eternal and uncreated, he would invoke a detached kind of deity to be responsible for the motions of the heavens, the "unmoved movers." These immaterial godlike entities had as their mission to impart the motions observed in the skies without themselves moving or either suffering or imposing any material cause. Mysteriously, they would impart motion through some sort of "aspiration or desire." Since Aristotle's cosmos was a complex onion-like contraption of spheres within spheres, with Earth static at the center and the sphere carrying the stars at the periphery, there was a vertical hierarchy within the unmoved movers, with the outermost one being the First Mover. Its role was to impart motion from the outside in, a kind of Cosmic Winder responsible for initiating the causal chain that animated the Universe.[4]

With his First Mover and subordinate collective of unmoved movers, Aristotle addressed two fundamental issues one faces when trying to explain Nature: how the change from rest into motion occurs, and how motion is maintained. How else could he explain what kept his huge cosmic machine in motion for all eternity? Aristotle missed the notion of inertia, the natural condition of a body to remain in its state of motion unless compelled to change it by a force. For that, some eighteen centuries had to pass.

Being eternal in time, Aristotle's cosmos had an added simplicity over a cosmos that appeared at some moment in the past, as in the biblical narrative or, more to the point, the modern Big Bang cosmology. As we remarked before, a Universe with a beginning needs a causal explanation to its appearance. Why should there be a Universe in the first place? What caused it to become? Religions resolve the issue by imposing the existence of a godlike First Cause that exists beyond the constraints of physical laws. To explain the emergence of the physical Universe through science is a tremendous conceptual challenge, one that still haunts modern cosmologists, even if many insist that quantum mechanics can take care of it all, which, as we shall argue later, is bad philosophy and scientifically fallacious. To claim that we know how the Universe emerged is both untrue and a great disservice to the public understanding of science. Like it or not, there is a horizon around every island. The Island of Knowledge is no exception.

Back to Aristotle. We see that he drove philosophy out of Plato's cave, dissolving the distinction between a world of abstract Forms and a world of sensorial perception. Change on Earth and its immediate surroundings is to be understood through the interplay of the four basic substances. As we ascend to the skies, we transition into the realm of the heavenly spheres, responsible for carrying the Moon and five planets in their circular orbits. (Mercury, Venus, Mars, Jupiter, and Saturn were the only planets known until the discovery of Uranus in 1781.) Heavenly objects, however, were made of a fifth essence, the perfect and eternal aether, immune to any kind of change. Aristotle's cosmos still retained a dualistic nature, a division between the terrestrial world of ordinary matter and the ethereal worlds above, inaccessible to matter and perfect. It also retained a version of an unknowable teleology, incorporated into the immaterial but active unmoved movers, a shoo-in for medieval Christian theology.

For the next few hundred years, different models attempted to explain the irregular motions of the heavenly objects within the Earth-centered Aristotelian cosmos. For as it was known since Sumerian times, the planets are not so nicely behaved: follow Mars for a few months across the sky, and you will notice that it sometimes appears to move backwards, as if unsure of the right way to go. This "retrograde" motion was a real headache to the ancient Greeks, who saw the cosmos from the standpoint of a central Earth. According to Simplicius of Cilicia, the sixth-century philosopher and commentator of Aristotle, Plato charged his disciples with explaining these irregular motions only with circles and uniform speeds, a challenge curiously known as "saving the phenomena." (Usually scientists don't save the phenomena but try to save theories from crumbling as they confront the phenomena.) Simplicius explained Plato's challenge this way: "And this is the wonderful problem for the astronomers: with certain hypotheses being given by them, to prove that all things in the heavens have a circular motion and that the apparent non-uniformity in each of them . . . is merely apparent and not real."[5]

Here we see how a powerful theoretical prejudice can work simultaneously as a mind closer and as a mind opener, jostling the imagination into creating viable scenarios within its rigid constraints.

Although Plato's dream led astronomy astray for almost two thousand years, it also triggered the invention of highly sophisticated explanations that attempted to accommodate the motions of the heavens within its short leash. Of these, Ptolemy's model of epicycles is the most prominent, having been proposed around 150 CE and surviving, with minor modifications and improvements from Islamic astronomers, until the late sixteenth century.

Briefly, an epicycle is a circle attached to a larger circle (called the "deferent"). Imagine the Earth at the center of the larger circle. Imagine now an epicycle attached to the large circle and the Moon attached to this epicycle. As the large circle turns, the epicycle turns with it. If, additionally, the epicycle could also turn on itself, we have a combination of two circular motions that, together, could generate a loopy path, simulating retrograde or irregular motion. Now repeat this procedure for each planet and the Sun: each sits on its own epicycle, which is attached to its own large circle. The whole construction would look like a series of concentric circles with the Earth in the center of them all. If the sizes of the large circles and their epicycles were adjusted, the amount of retrograde motion could be made to vary so as to match astronomical observations for the different celestial objects.

Ptolemy quickly realized that such a simple construction would not work. Searching for the most precise model possible, capable of predicting the positions of the celestial objects far into the future, he added an extra feature to his celestial machine: instead of turning regularly around the center of the large circle like in a Ferris wheel, the epicycles would turn around an imaginary point slightly displaced from the center of the large circle called the "equant." With this modification, Ptolemy's model had the remarkable accuracy of predicting the positions of the planets with a precision roughly equivalent to that of a full Moon. That is, his prediction would have an error smaller than the space the full Moon occupies in the sky.

Ptolemy and most of his Islamic followers never believed the epicycles were real. To them, they were mere calculating devices used to predict the positions of the various celestial objects. Moses Maimonides, the medieval Jewish Aristotelian, writes clearly about this after he argues against the physical existence of epicycles: "All this does not affect the astronomer. For his purpose is not to tell us in which way the spheres truly are, but to posit an astronomical system in which it would be possible for the motions to be circular and uniform and to correspond to what is apprehended through sight, regardless of whether or not things are thus in fact."[6] In other words, although contemplation of the heavenly motions might bring humans closer to God, astronomy is not concerned with explanations about the nature of things, only with describing the motions of things celestial, which are "apprehended through sight," that is, through observations. There are thus things that can be understood—those apprehended through the senses—and things that cannot be understood—those that go beyond the sensorial. Maimonides then acknowledges that the true reality of the heavens is unknowable to humankind:

> For it is impossible for us to accede to the points starting from which conclusions may be drawn about the heavens; for the latter are too far away from us and too high in place and in rank. And even the general conclusion that may be drawn from them, namely, that they prove the existence of their Mover, is a matter the knowledge of which cannot be reached by human intellects. And to fatigue the minds with notions that cannot be grasped by them and for the grasp of which they have no instrument, is a defect in one's inborn disposition or some sort of temptation.

Of course, we have learned much about the nature of the heavens since the times of Maimonides. But it would be foolish to quickly dismiss his words as nonsense or as defeatist. For we must recognize that because of the very nature of human inquiry every age has its unknowables. The question we need to address, then, is whether certain unknowables are here to stay or whether they can be dealt with in due course. Must every question have an answer?

While the epicycles weren't real, the crystal spheres carrying the celestial objects around were believed to be. Perhaps no idea in the history of astronomy has had such longevity. The first mention of a crystal ("ice-like") sphere in the cosmos is attributed to Anaximander's disciple Anaximenes, another of the Ionian Presocratics from Miletus. According to Aëtius, "Anaximenes says the stars are fixed like nails to an 'ice-like' surface so as to form designs."[7] Although there is some debate as to whether it was Anaximenes or Empedocles who first suggested the notion of revolving orbs carrying celestial objects around the Earth, it is clear that by Plato's time and, in particular, in the nested spheres model of his disciple Eudoxus of Cnidos, revolving spheres were the essence of the cosmic machinery. Even Copernicus, some eighteen centuries later, was still convinced that crystal spheres carried the planets along their orbits. The groundbreaking book in which he proposed that the Sun, not the Earth, occupied the center of the cosmos was entitled *On the Revolutions of the Heavenly Spheres*. Indeed, without the steadily revolving spheres, how would one explain the motion of the celestial objects? The only understanding of gravity came from Aristotle, who divided the cosmos into two realms, each with its own "physics." And he used no fewer than fifty-nine spheres to keep the heavens in check. Copernicus recognized he had a challenge but didn't know how to fully address it.

By moving the Sun to the center, Copernicus precipitated a major cosmic cataclysm, an upheaval of the Aristotelian order that had ruled for almost two millennia. The new world order required new explanations, a new science Copernicus was unable to provide. Under Aristotelian physics, a central Earth was the attractor point of all material motions, the reason why things fell to the ground. In turn, celestial orbs carried the Moon, Sun, planets, and stars about the center in regular, or at least epicycle-driven, circular motions. With the Earth relegated to a planet, why should things fall down? To complicate things, the Sun and all celestial objects were supposedly made of a fifth essence, the aether, of a completely different nature from the four elements found down here. The aether was eternal and immutable: nothing changed in the skies. Even asteroids and comets the Aristotelians attributed to atmospheric or "meteorological" phenomena.[8] Why would the Earth, not made of aether, be on equal footing with the aethereal planets? What kind of physics could make sense of this mess?

Theologically too there were problems. The new planetary arrangement meant removing the natural verticality of the Aristotelian cosmos that the church embraced with such enthusiasm, a verticality that made people look up in awe toward the heavens—the realm of God and the chosen. In addition, if the Earth circled the Sun, hell would not be at the center of Creation any longer, but revolving with the whole Earth about the skies. No wonder Martin Luther was one of the first to condemn Copernicus: "There is talk of a new astrologer who wants to prove that the Earth moves and goes around instead of the sky. . . . The fool wants to turn the whole art of astronomy upside-down."[9]

Copernicus didn't want a revolution. He so wanted to go back to Plato and "save the phenomena" that he proposed a cosmos based on the rule of beauty and symmetry, anchored once again on regular circular motions. He despised Ptolemy's equant, as it imposed an irregularity on the celestial motions. A man of the Renaissance, who studied in Italy just a few years before Michelangelo painted the Sistine Chapel, Copernicus believed the Sun-centered cosmos offered a "bond of harmony," a new cosmic aesthetic missing in the age-old, Earth-centered alternative. His system of the world paid homage to Philolaus and the Pythagorean notion that a central fire had to be the cornerstone of Creation, from whence all light emanates. Copernicus's model was Plato's dream couched in the values of the Renaissance, a cosmos built around the idea of beauty and symmetry with little input from new quantitative astronomical observations. Indeed, Copernicus made only a small number of original astronomical observations, using mostly data from Ptolemy and his Islamic successors.

The key difference between Copernicus and his predecessors was the attitude toward the reality of his vision: to Copernicus, the Sun-centered cosmos was not simply a computing device but the true arrangement of the world. Astronomy was not a mere description of the cosmos but a mirror of physical reality as perceived by the human mind. The stakes suddenly became much higher.

However, the watershed was only to begin in earnest six decades after the publication of Copernicus's book, mostly because of Galileo and Kepler. To both men, the key factor was a revolutionary influx of new observational evidence: Galileo's life and the future of astronomy changed after he put his hands on a telescope, and Kepler's groundbreaking physical astronomy would not have been possible without Tycho Brahe's meticulous data.

CHAPTER 5

THE TRANSFORMATIVE POWER
OF A NEW OBSERVATIONAL TOOL

*(Wherein we describe how three remarkable gentlemen,
with access to new observational tools and endowed with
remarkable creativity, transformed our worldview)*

Before a Dutch-made telescope reached Galileo's hands in the fall of 1608, Tycho Brahe had spent the last three decades of the sixteenth century measuring with great care the motions of the planets in the skies. Thanks to his personal wealth, and with the support of King Frederick II of Denmark, who in 1576 awarded him the island of Hven—"with all the crown's tenants and servants who thereon live, with all rent and duty which comes from that . . . [for] as long as he lives and likes to continue and follow his *studia mathematices*"[1]—Tycho built an ensemble of measuring instruments the likes of which the world had never seen. In these pretelescopic days, astronomical measurements were performed exclusively with the naked eye, relying on quadrants, sextants, astrolabes, and other instruments to measure the locations and motions of the celestial bodies. These translated into angular measurements along the celestial sphere, the imaginary dome that the stars appear to be attached to.

Stepping outside on a clear, cloudless night, we see a multitude of stars (a few thousand of them at most), which seem to maintain their mutual distances as if indeed nailed to the darkness. The whole celestial dome moves slowly east to west as the night unfolds. The apparent relative immobility of the stars inspired ancient sky gazers to attach meaning to what seemed to be drawings in the skies: the constellations. Although different local mythologies would attach different meanings to the same constellations, the impulse to extract messages from the night sky is a very pervasive feature of human culture. In reality, and illustrating how our senses deceive us, the stars are neither static—some move with speeds of thousands of miles per hour—nor on the same two-dimensional celestial dome, being at varying distances from Earth and hence spread across the vast three-dimensional volume of interstellar space. The celestial dome is like Plato's cave wall, an illusion resulting from our limited perception of reality. (But in this case, supposedly without a conjuror behind the scenes.) The illusion has to do with the enormous distances separating us from the stars. Just as an airplane flying high in the sky appears to move slowly for someone on the ground, visible stars, with distances of, say, a few to hundreds of light-years from us, appear static.[2]

If you live in the Northern Hemisphere, a long-exposure photograph reveals that the sky seems to rotate around a single star, Polaris, the North Star. In fact, it is the Earth that is rotating and not the sky: Polaris happens to be aligned (for now) with the Earth's North Pole. This alignment will gradually drift away within the next few thousand years because of Earth's slow wobbling about the vertical like a falling top, a motion called "precession of the equinoxes."

The very counterintuitive notion that the Earth rotates about itself eluded human observers for thousands of years. To the suggestion that the Earth, and not the skies, turned about itself in twenty-four hours Aristotelians would respond that if the Earth rotated, clouds and birds would be left behind, as would stones thrown up in the air. With the exception of a few Greek thinkers such as Ecphantus and Heraclides, the outrageous notion of a rotating Earth would only be firmly suggested again by Copernicus almost two thousand years afterwards.

To measure the relative positions of stars and the moving planets in the sky, astronomers divide the celestial sphere into two hemispheres, bisected by the Earth's equator. Polaris is at the top (or zenith) of the Northern Hemisphere. The elevation above or below the celestial equator is called the "declination" (the same as the latitude on Earth's surface). So, Polaris is at declination of +90 degrees. In analogy with the terrestrial longitude, which is measured from the arbitrary zero point fixed at Greenwich, England, the position along the circle of the celestial equator is the "right ascension." Conventionally, the zero-point of right ascension is chosen at the vernal equinox (the beginning of spring), when the Sun crosses the celestial equator.[3] A slight complication is that instead of angles, as is the case with latitude and longitude (and declination), the right ascension is measured in hours, minutes, and seconds. To connect the two angular units astronomers use the rotation of the Earth. Since Earth rotates around itself by 360 degrees (a full circle) in twenty-four hours, in one hour it rotates by $360°/24 = 15°$; in one minute by $15°/60 = 0.25°$ (aka 15 arc minutes or 15'); and in one second by $15'/60 = 15''$ (aka 15 arc seconds). So a right ascension with angular position of 15 degrees with respect to the zero point is expressed as 1^h, for one hour. As an example, to find the star Betelgeuse from the Orion constellation on the celestial sphere, look $5^h52^m0^s$ east of the vernal equinox (its right ascension) and $7°24'$ north of the celestial equator (its declination).

Back to Tycho Brahe. His large, built-to-order instruments allowed him to measure planetary positions with the unprecedented precision of 8 arc minutes of a degree.[4] Tycho also understood that to resolve the shapes of the planetary orbits, precision had to be combined with regularity: the more data points he had, the more precisely he could trace the motions of the planets across the sky. And so it was that, on November 11, 1572, while strolling back from his alchemical lab, Tycho saw a new star in the constellation of Cassiopeia. The astonishing apparition was so intense that it could be seen in plain daylight. According to Aristotelian physics, a new celestial luminary was an impossibility: the skies were immutable, and changes only occurred below the lunar sphere. Any new celestial phenomenon was deemed a mere atmospheric disturbance, belonging to the study of "meteorology." Armed with his instruments, Tycho meticulously measured the new celestial light until it faded from view in March 1574. His conclusions were revolutionary: first, the "new star" was farther away than the Moon; second, it wasn't a comet, as it didn't have a tail or move across the sky. Tycho's results were the first serious observational challenge to the Aristotelian credo. It takes no small dose of intellectual courage to proclaim that the established order is incorrect, that changes are afoot. Tycho's game-changing modernity was evident in his pursuit of high-level accuracy in measurement and in his understanding that theories without support from data are like empty shells, beautiful to behold but lacking the substance that gives them their raison d'être.

We now know that Tycho had spotted a supernova event, the explosive death of a large star. What he thought was a new star in the sky was actually an old star dying out. Tycho's magnificent instruments and diligence allowed him to see more clearly than anyone before him; nevertheless, as is often the case in the history of science and central to our argument, his vision was blurred by how little he could see. The accusatory cry he unleashed on those who doubted him could be uttered to all of us: "Oh thick wits. Oh blind watchers of the skies."

As if the skies were spurring things on, in 1577 another apparition fed more wood to the slowly spreading anti-Aristotelian fire: the Great Comet of 1577, visible throughout Western Europe and recorded by many astronomers. Tycho spotted it coming back from a fishing trip just before sunset on November 13, and followed it for seventy-four days.[5] Comparing his data with that of an astronomer from Prague, Tycho concluded that the comet was at least three times farther away than the Moon: he noticed that while the Moon was in a different place for his colleague, the comet was not. This technique is know as parallax and is extremely useful when trying to establish the relative distance between faraway objects.[6] Other astronomers confirmed Tycho's finding, further undermining the basic Aristotelian precept of celestial immutability.

Considering that Tycho's discoveries happened three decades after the publication of Copernicus's book in 1543, one would imagine that he would have eagerly embraced the heliocentric model. Alas, he did not. For physical and theological reasons he remained unconvinced, proposing instead a weird hybrid model with two centers: the Earth continued at rest at the dead center of Creation with the Moon and Sun revolving around it, but the rest of the planets circled the Sun. Tycho's lopsided model was a result of both his upholding of the biblical narrative and a steadfast belief in the power of observations. He carefully mapped and compared the relative positions of stars at different times of the year for any indication that the Earth moved (applying to stars the same parallax technique he used on the Great Comet of 1577) but found nothing. If the Earth moved about the Sun, at different times of the year nearby stars would appear to be at different positions relative to distant ones. Tycho found nothing because it is impossible to detect stellar parallax with the naked eye, even with instruments as good as his. He, like everyone else, was a "blind" watcher of the sky, albeit one who could see farther than most. Tycho also couldn't see what kind of new physics could justify a Sun-centered model. Even if his observations pointed away from the age-old Aristotelian division of reality into two separate realms, Tycho wasn't prepared to take the next logical step and risk the possibility that a whole new physics lay out there, waiting to be discovered.

Still, he was bold enough to do away with crystal spheres, as some would clearly overlap with each other in his model. He figured that comets would presumably fly right through them like fiery bullets, leaving only broken shards in their trail, although his data didn't quite have the accuracy to prove it. Getting rid of the sacrosanct crystal spheres presented Tycho with a huge challenge: explaining the motions of the celestial objects if spheres were not there to carry them along. Supremely confident of his data, Tycho chose to leave the planets hovering in empty space, unsure of how to explain the celestial revolutions he so carefully measured. He needed an architect, someone with the vision and mathematical stamina to show that his model reproduced the true cosmic arrangement. He needed Johannes Kepler.

Few characters in the history of science are as fascinating as the brilliant, neurotically intense, and courageous German astronomer who, in his darker moments, saw himself as nothing more than a puny lapdog, while being, in fact, an intellectual giant and a hero of the fight for religious freedom. Emotionally scarred from the tragedies of a highly dysfunctional upbringing and personal life, a victim of the virulent religious strife between Catholics and Lutherans that ravaged Central Europe during the first decades of the seventeenth century, Kepler turned his mind to the skies in search of the order his life stubbornly refused to give him.[7]

Kepler joined Tycho as his assistant early in 1600. By then, the astronomer-prince had fallen from grace with the Danish crown and had become imperial mathematician to Rudolph II, the Holy Roman Emperor who ruled from Prague. Tycho, in keeping with the splendor and excess of his previous life, had built a sophisticated astronomical observatory for himself and his entourage of assistants and instruments at Benatky Castle, about forty miles outside of Prague.

From the start, Tycho and Kepler had very different agendas. Tycho wanted a theoretician to justify the lopsided, Earth-centered model that he believed was consistent with his observations and with the Scriptures. Kepler, a devoted Copernican, wanted to use Tycho's data to establish once and for all the veracity of the Sun-centered cosmos. Although they worked together only for eighteen months, their clash was epic. Tycho was not about to hand his life-long labor, decades of meticulous work, to the German Copernican. Kepler, on the other hand, couldn't wait to get going. After much give and take, Tycho finally gave Kepler his data on the motion of Mars. It was a sneaky move on Tycho's part. He knew Mars has a relatively high eccentric orbit, meaning that it departs quite sharply from a simple circle.[8] Kepler's task was to explain Mars's imperfect orbit with circular motions, consistent with Tycho's data.

What Kepler thought would take him a couple of weeks took him almost nine years. In 1609, he proudly published his "New Astronomy," where he declared that the orbit of Mars was not a circle but an ellipse. To arrive at such a mind-boggling conclusion contradicting thousands of years of astronomy, Kepler held tight to Tycho's data. After years of trying all sorts of variations around circles and epicycles, Kepler applied Ptolemy's idea of an equant to the Sun, slightly displacing it from the center of all planetary orbits. It almost worked, but for two measurements that disagreed with his model's predictions by 8 minutes of arc, that is, by 8/60th of 1 degree. Most people would have called it quits at this point, declaring the model to be a very good approximation to the data—which it was. But relentless Kepler moved on, knowing that he could do better. Only then would he honor the precious data he had in his hands.

And so Kepler kept on trying, until he hit on the ellipse. It was the second time he had tried it, having given up on it earlier; sometimes the answer is staring you in the face, but you are not ready to embrace it. The agreement was spectacular. In Kepler's hands, Tycho's meticulous data, obtained with high-precision naked-eye astronomy, would spark a revolution in human knowledge. Few examples in the history of science illustrate so clearly the power of high-precision data as a catalyst for a revolutionary shift in our collective worldview. In Tycho and Kepler's story we witness how formidable the alliance between observer and theorist can be, even if not always so dramatic. Paraphrasing Einstein's famous dictum on science and religion, "Data without theory is lame; theory without data is blind."

Kepler didn't stop there. To be the agent of change, he had to go beyond "merely" justifying the Copernican astronomy with Tycho's data; he had to provide a new physics to explain it. The subtitle of his book says it all: *A new astronomy based on causation or a physics of the sky derived from the investigations of the motions of the star Mars founded on observations of the noble Tycho Brahe*. A new astronomy based on "causation or a physics of the sky." Not only was Kepler seeking a descriptive astronomy, as had been all his antecessors, but he wanted to explain astronomy through physics, convinced that the motions of the planets had a causation mechanism. This is truly revolutionary, the first attempt in the history of astronomy to include a law of cause and effect, to explain planetary orbits as the result of a force. Kepler suggested that the Sun and planets have magnetic natures and that their mutual interaction is through magnetism. His inspiration was a book by William Gilbert, Elizabeth I's court physician, who described the Earth as a giant lodestone. If the Earth was a magnet, reasoned Kepler, the Sun would be one too. And as two magnets attract each other through empty space, so would the Sun and planets, even if separated by great distances. As he wrote in a letter from 1605, "my aim is to show that the celestial machine is to be likened not to a divine organism but rather to a clockwork . . . insofar as nearly all manifold movements are carried out by means of a single, quite simple magnetic force." Kepler's groundbreaking ideas on a physical cause for the planetary motions would be the cornerstone of Newton's theory of gravity, developed later in the seventeenth century.

Before leaving Kepler, I will point out a second instance of his unprecedented modernity in his book's subtitle: . . . *a physics of the sky . . . founded on observations* . . . Often taken by arresting flights of fancy in his speculations about the cosmos, Kepler also knew that data was the final arbiter between Nature and the theories we build to make sense of it. Although this fact may seem obvious to us now, it wasn't at all then. Kepler was a transitional figure, a harbinger of things to come. But by then he wasn't acting alone. In distant Italy, another Copernican was ready to come out of the closet.

In 1610, just a year after Kepler's *New Astronomy*, Galileo published his *Siderius Nuncius*, usually translated as *The Sidereal Messenger*. In this little book, Galileo set off to change the way we look at the Universe. He could do so because he had a powerful new tool to look at the skies: a telescope. And what he saw revealed a cosmos of awesome beauty and complexity, a new kind of beauty far from the idealized Aristotelian perfection of ethereal spheres in an unchangeable heaven. Just as Tycho's instruments allowed him to measure the skies with unprecedented accuracy, Galileo's telescope allowed him to see farther and clearer than anyone before him. As frequently happens in the history of science, a new observational tool revealed new and often unpredictable aspects of physical reality. The Island of Knowledge grows in fits and starts, as new land redefines its coastline and pushes older bits inland or into oblivion.

Although news of a telescope was circulating as early as October 1608, when the Dutch lens maker Hans Lipperhey filed for a patent to secure the rights of the invention (it was denied), Galileo's first telescope was his own creation. Aware of the enormous potential of this new instrument, Galileo decided to improve on an exemplar a diplomat friend had given him, grinding his own lenses to assemble a telescope with a three-times magnification by July 1609. Shortly after, he presented an eight-times magnification instrument to the Venetian senate in August—a move that secured his position at the University of Padua and doubled his salary—and, in October, turned a twenty-times magnification telescope to the skies. Galileo was not alone. We now know that Thomas Harriot of England observed the Moon with a six-powered instrument in August 1609, although he published nothing of his findings.[9] The telescope owes its fame to Galileo and his tireless efforts to promote it as a new way of doing astronomy, the tool that would precipitate a new world order.

Much has been written about Galileo and his trials and tribulations with the Catholic Church, including in my own *The Dancing Universe*. Here our focus will be on the impact of his discoveries and his key role in developing the empirical methodology that would become the trademark of modern science.

In *The Sidereal Messenger* (no doubt Galileo saw himself in this prophetic role), Galileo describes three main discoveries made with his telescope, all opposing the Aristotelian view of the cosmos: the Moon's surface, far from being perfect, displayed mountains and craters, being thus more similar to Earth than to some aether-like spotless sphere; pointing his telescope to the Pleiades and Orion constellations, he saw at least ten times more stars than with the naked eye, inferring that the Milky Way and other nebulae were no cloud-like objects but collections of countless stars; Jupiter had four satellites, which Galileo astutely baptized "Medicean stars," vying for the patronage of Cosimo II de' Medici, the Grand Duke of Tuscany. These discoveries and others he made later on, such as the phases of Venus and the existence of sunspots, convinced Galileo that Copernicus was correct and Aristotle wrong.[10] Even if they didn't quite constitute proof of the Copernican hypothesis (Tycho's model could explain them all) as say, stellar parallax would, Galileo decided to tell the world and the church that it was time to change, eventually bringing the wrath of the Inquisition upon him.

For all of its modernity and revolutionary fire, Galileo's astronomical work still showed marked conservatism. In particular, he never believed in Kepler's elliptical orbits. Instead, he adapted ideas by the fourteenth-century Oxford scholar Jean Buridan to propose a strange law of circular inertia to justify the circular motions of planets around the Sun. He then extrapolated this law to a law of linear inertia: "A body moving on a level surface will continue in the same direction at a constant speed unless disturbed." (Just think of an ice skater gliding along the smooth flat surface of a frozen lake.) Newton would later adapt this law into his first law of motion, although in his version the concept of force makes an entry: "Unless acted upon by a net unbalanced force, an object will maintain a constant velocity." By the way, the word "inertia" makes its first appearance in Kepler's *Epitome Astronomiae Copernicanae*, published in three parts between 1618 and 1621. In this masterpiece of early modern astronomy, Kepler generalizes his elliptical orbits to all planets and successfully tests his mathematical formulas using Tycho's data. Inertia, to him, represented a body's resistance to be set into motion, starting from rest.

Still, to Galileo and to Kepler the cosmos remained closed, contained within the sphere of the fixed stars. Kepler, in fact, believed that an infinite universe was an abomination: "This very cogitation carries with it I don't know what secret, hidden horror; indeed one finds oneself wondering in this immensity to which are denied limits and center and therefore also all determinate places."[11]

Kepler believed that a cosmos created by God would have geometric order and symmetry and not be endless and formless. He even equated the cosmos with the Holy Trinity: the Sun, at the center, is the Father; the Son, at the periphery, the sphere of the fixed stars; and the intervening space between the two, filled with the Sun's (God's) light, the Holy Spirit. To buttress his theological argument, he maintained that an infinite universe contradicted observations, citing the supernova of 1604 as an example ("Kepler's supernova," the last to be seen with the naked eye). Defenders of an infinite cosmos argued that the new star became visible as it descended from the depths of space and faded away as it ascended back into oblivion. Kepler denied this notion, claiming that stars don't move. He further argued that an infinite cosmos would be homogenous, looking the same at all places, which it clearly wasn't, as one could see from the disposition of the stars in constellations.

It is also possible that both Kepler and especially Galileo had in mind Giordano Bruno's horrible fate in the hands of the

Inquisition, even if Bruno's condemnation and public burning resulted more from his theological iconoclasm than from his astronomy. For example, he argued that Jesus wasn't the son of God but a skillful magician and that the Holy Ghost is the soul of the world. However, his belief in the infinitude of the cosmos and that each star is a sun with its own court of planets (how right he was!) and their own thinking beings also clashed with the centrality of the Earth and with the belief that humans are God's favored creatures.

After Galileo and Kepler had set the stage, the next major shift in the conception of reality happened in the hands of Isaac Newton, who not only developed a precise formulation of the law of gravity applicable to any object in the Universe, but also cracked open the dome of heaven, arguing for the infinity of space. No single mind had caused such an enormous expansion of the Island of Knowledge. And very few would afterwards.

CHAPTER 6

CRACKING OPEN
THE DOME OF HEAVEN

*(Wherein we explore the genius of Isaac Newton and why
his physics became a beacon of the human intellect)*

Galileo died in 1642, the same year Newton was born. The Italian iconoclast didn't restrict his work to the nature of the heavens. Down on Earth too, he shook the very foundations of Aristotelian physics, showing, to the surprise of countless souls and the disgust of many church fathers, that appearances indeed were deceiving. The most spectacular of Galileo's discoveries pertains to the nature of gravity. Even today, when I lecture on how things fall and show how wrong our intuition is, I see surprise and often resistance stamped on my students' faces. As Aristotle would have it, and the senses seem to confirm, an object moves innately toward its "natural place." The "natural places" were organized hierarchically according to the four basic elements. From bottom up: earth, water, air, and fire. Makes perfect sense, really, given that we see a rock fall to the ground when suspended in air (or water or fire), while we see fire slicing upwards through air. A corollary of this arrangement is that heavier objects fall faster: gravity would, as it were, respond to the constitution of an object. And why not, given that we see a feather fall to the ground much more slowly than a rock?

In a series of amazing experiments, Galileo showed that Aristotle and our intuition were simply wrong. All objects, irrespective of weight, shape, or constitution, fall down at exactly the same rate. The differences can be attributed mostly to air friction or small differences in the release time of the two objects. More precisely, we should say that all objects, irrespective of their mass, fall down at the same rate in a vacuum, although the distinction between weight and mass would have to wait for Newton. Galileo described the kinematics of free fall, measuring how fast different objects fell. To quantify motion and time, he had a brilliant idea: he let balls roll down along inclined planes, varying the angle of inclination to control their speed. This way he could count how long it would take for the ball to roll down at different inclinations, even before the invention of clocks: to measure time he used his pulse, music (as humans have a remarkable ability to keep a beat), and even water pouring into a bucket. To make sure the Aristotelian coffin was duly nailed shut, Galileo performed two more experiments. One was the legendary dropping of wooden and lead balls from the top of the Tower of Pisa; despite their very different weights they hit the ground nearly together.[1]

The other experiment with falling bodies happened earlier, during mass at the Pisa cathedral in 1602, when Galileo's attention drifted to the altar boy lighting up the candles in a large candelabrum. Galileo noticed that after the altar boy let go, the candelabrum oscillated back and forth. To his amazement, even if the swinging gradually diminished in amplitude, the time between full swings (the period of oscillation) remained approximately the same. (This is only strictly true for small amplitude swings.) Afterwards, Galileo showed that the period was independent of the mass of the object: when released from the same position (same angle with respect to the vertical), heavy bobs or light ones oscillate with the same period. What determines the period of a pendulum (for small-amplitude oscillations) is only the length of the string and the local value of the force of gravity (which was constant for Galileo's experiments).

Given that pendular motion is essentially controlled falling, the fact that different masses had the same period of oscillation was consistent with balls rolling down inclined planes and the Tower of Pisa experiments: falling down is a democratic exercise, where all masses are treated equally. Differences that are seen, say, between a feather and a Cadillac dropped from ten feet high, are due to air friction. At the end of his Apollo 15 moonwalk, Commander David Scott dropped a hammer and a feather to test Galileo's results in the absence of air. The video is striking and, albeit not surprising to those who know about Galileo's findings, still feels like magic.[2] The wonderful thing is that it isn't.

While Kepler was obtaining the first mathematical laws describing astronomical orbits, Galileo was obtaining the first mathematical laws describing motions close to Earth. Nature became amenable to rational exploration, through the combination of mathematical formulation and data gathering. Both Kepler and Galileo reached their results after much careful experimenting and toying with data, obtaining what we call empirical natural laws. Among other things, their story teaches us that experimental accuracy is crucial to uncover mathematical order in Nature: Kepler with his 8 arc minutes and Galileo with his free-fall timing measurements. The physical sciences require a methodology that employs equations and high-precision instruments. A measurement is a number, and a sequence of measurements may denote a trend. The role of the scientist is to coax meaning out of this trend, exploring plausible regularities and expressing them in terms of a mathematical law applicable to systems of the same nature. Kepler's laws of planetary motion work for any orbiting object in this or another stellar system (so long as gravity is not too strong), and Galileo's results on free-falling objects are valid for any (constant) gravitational field.

Newton comes in as the great unifier, the man who ties the physics of the Earth with that of the heavens. He shows that both Galileo's law of free fall and Kepler's laws of planetary motion are, in effect, the same, as he expressed in his law of universal gravity. Newton brought the heavens nearer to Earth and nearer to us, allowing human minds to scrutinize its mysteries. If the empirical laws of his predecessors were an expression of regularity in the sky and on Earth, his law revealed a cosmic order without precedent in the history of thought. Newton, a devoted alchemist, must surely have reveled in giving precise expression to the famous dictum from the *Emerald Tablet* of Hermes Trismegistus, considered the essential code of alchemy: "That which is above is that which is below."[3] To Newton, the mathematical principles of natural philosophy, the alchemical search for unity of matter and spirit, and God's role as Creator and keeper of universal order were deeply related.

The mechanisms of the cosmic clockwork, from the most distant planet to the falling apple, obey a strict set of rules, expressed in a single equation. No wonder Newton is celebrated as the grand architect of modern science, the man who more than anyone before him embodied the power of reason to make sense of the world around us.

What many forget is that Newton wasn't your typical lonely theoretician, searching for mathematical laws of Nature in his cloistered Cambridge lodgings. That he was a recluse, adverse to direct social or intellectual exchange, is well known and documented in many biographies. What is perhaps less known is that Newton was also a careful experimentalist, who devoted long hours to exploring the optical properties of light and, with even more fervor, the alchemical search for hermetic enlightenment. We will return to the latter soon enough.

In optics, Newton elucidated the nature of visible light—that it is composed of a superposition of (infinitely) many colors, bracketed between the violet and the red of the rainbow—and invented a new type of telescope: a reflector, far superior to the refractor

telescope of Galileo in that it provided images with higher resolution and free from annoying color distortions known as "ab-
errations." Newton's reflector telescope, using a curved mirror to
gather light and focus it into the eye of the observer, catapulted him
to fame even before his laws of motion and universal gravity. By
1669 Newton had already become the second Lucasian professor
of mathematics at Cambridge University, a chair founded in 1663
and still going strong. Stephen Hawking held it from 1979 until
2009, and Michael Green, a noted string theorist, has held it since
Hawking's retirement.

In December 1671, the first Lucasian professor and Newton's
admirer Isaac Barrow took the reflector telescope to London, to
share with members of the Royal Society, the famous learned society devoted to improving the knowledge of Nature. A month later,
Newton joined as a fellow, sealing his inclusion into England's scientific elite. With fame, however, came exposure, and with exposure came professional envy and intellectual confrontation, games
Newton wasn't willing to play, at least initially. Only after the 1687
publication of his masterwork *Principia*, in which he presented the
laws of motion and gravitation, and the deserved acclaim as one
of science's all-time greats, would Newton venture once more into
the public sphere.

With the exception of a chosen few, including the pioneering
chemist Robert Boyle, Newton kept his alchemical-related work
mostly to himself. (He reserved the same secrecy for his theological
studies.) His new theory of the world, however, affected every realm
of knowledge, spreading like brush fire far beyond his control. A
theory that explains the celestial dynamics through the action of invisible forces would necessarily attract the interest of theologians—
even more so if those forces stretch across the cosmos, from the
falling of the smallest object to the ground to the majestic motions
of comets and planets. How could a believer in an all-powerful
Creator not see God's will in gravity's invisible threads? Accordingly,
as Newton explained to Cambridge theologian Richard Bentley,

only an infinite cosmos would befit God's infinite creative power. Indeed, given that God is in all of space, space must be infinite. In the General Scholium of the *Principia*, Newton expressed his belief that God and the Universe were one and the same: "[God] endures always and is present everywhere, and by existing always and everywhere he constitutes duration and space."[4]

Newton's new theory of gravity cracked open the dome of heaven, extending space to the furthermost reaches of infinity. It is a vision of immense beauty and awe-inspiring terror, a cosmos of countless suns "placed at immense distances from one another," the precarious Earth but a speck in an infinite expanse devoid of a center, of a comforting place where we belong. Decades before Newton's revolutionary ideas met the public eye, the French mathematician and philosopher Blaise Pascal had echoed Kepler in anticipating the existential angst of an infinite world of long (eternal?) duration: "The eternal silence of these infinite spaces frightens me." Or, more in depth,

> When I consider the short duration of my life, swallowed up in the eternity before and after the little space that I fill, and even can see, engulfed in the infinite immensity of spaces of which I am ignorant, and which know me not, I am frightened, and am astonished at being here rather than there, why now rather than then. Who has put me here? By whose order and direction have this place and time been allotted to me?[5]

Pascal's terror reflects the reaction many still have when faced with the ongoing revelations of science, which, in the three centuries following his words, have confirmed to extraordinary precision the true vastness of space and time. If not through religion, as Pascal's defense for Christianity would have it, how else would someone find meaning in our fleeting existence?

CHAPTER 7

SCIENCE AS
NATURE'S GRAND NARRATIVE

(Wherein we argue that science is a human construction
powerful in its range and its openness to change)

Newton and Galileo and Kepler found this meaning in understanding the workings of Nature. Many of those who followed them did as well. If, indeed, God was the creator of the world and its laws, then it was incumbent on the devout to search for these laws and, in so doing, unveil the cosmic blueprint: the grandest aspiration of the human mind, armed with mathematics, intuition, and high-precision data, was to decipher the divine plan. Even today, scientists who are also believers reconcile their faith and their science along these terms, claiming that the more they learn about Nature, the more they admire God's handiwork. And even among those who are not believers, many succumb to the notion of Oneness, of the unity of Nature.

We have seen how the game changed with Galileo, Kepler, and Newton, how science became increasingly tool-driven and how the limits of what we could know of the world were reflected in the efficiency of those tools. The regularities of Nature were to be expressed in terms of mathematical laws, obtained from painstaking observations of natural phenomena. With every discovery, the

Island grew broader, but so did the unknowns, the new questions that scientists were able to ask.

So efficient was the undertaking that by 1827, within one hundred years of Newton's death, scientific knowledge had been deeply transformed. Concepts such as energy and conservation laws were now recognized as being part of Nature's narrative, as were electric currents and magnetism. As ever more powerful telescopes scrutinized the heavens, physics had expanded its reach. The known planets had become seven, after William Herschel's discovery of Uranus in 1781; more comets were visible in their fiery orbits across the skies, while nebulae revealed amazing features of light and color behind their apparent cloudy demeanor. The cosmos was far richer and vibrant than anyone could have anticipated or dreamt. Ancient Ionian visions of a Universe in a constant state of flux came back to the fore. Inevitably, so did the opposing notions of perfection and immutability. To make sense of the cosmic narrative, it seemed that a balance would have to be struck between notions of symmetry, beauty, and conservation laws on the one hand and concepts of change, decay, and rebirth on the other.

As the cumulative knowledge of the world expanded, so did the breadth of our ignorance. Tools allowed for the amplification of our myopic view, revealing unanticipated richness at all levels from small to large. Theories, when successful, may anticipate new objects and properties of Nature, but they can't predict all that is out there. As new tools amplify our view of the world, they also reveal how much we don't know and couldn't have anticipated, often in spectacular fashion. Case in point: in the world of the very small, a twin revolution had been unleashed when the Dutchmen Zacharias Janssen and then Anton van Leeuwenhoek invented and perfected the microscope at around the same time that Galileo was pointing his telescope to the sky. Van Leeuwenhoek, in particular, upon examining scrapings off his teeth and finding bacteria, revealed a whole new world of microscopic life.

The discovery of tiny life forms prompted a whole new slew of questions: How far down could life go in size? Is there a difference between living and nonliving matter? How did life originate? Essential questions of the very large, such as the extension of space and the duration of the world, found their counterpart at the opposite extreme: How small could matter be? How long could it live? Is our mortality predetermined by God, or is it a natural occurrence? That inanimate matter could be animated into life without the mediation of anything divine was a mind-boggling possibility, one that scared many believers. It brings to mind Newton's fourth letter to Richard Bentley, in which the natural philosopher responded to the theologian's invitation to comment on the nature of gravity:

> Tis unconceivable that inanimate brute matter should (without the mediation of something else which is not material) operate upon and affect other matter without mutual contact. . . . That gravity should be innate inherent and essential to matter . . . is to me so great an absurdity that I believe no man who has in philosophical matters any competent faculty of thinking can fall into it.[1]

Newton was arguing quite forcibly that gravity cannot have a material explanation, since inert matter is, well, inert. There was something intangible within matter that somehow gave rise to gravitational attraction. Newton must have thought that the hand of God was part of this, although he was careful (and somewhat contradictory) in his final remark to Bentley on the subject: "Gravity must be caused by an agent acting constantly according to certain laws, but whether this agent be material or immaterial is a question I have left to the consideration of my readers."

Since Newton, forces became pragmatic explanations for the behaviors of material bodies, the hows that dictate the ways in which we capture the world through an application of our senses and their tool-aided extensions. There is no room for metaphysical speculation in "experimental philosophy," to use Newton's term, where "whatever is not deduced from the phenomena . . . ha[s] no place."[2]

This is, to this day, the credo of science. In an ontological description of the natural world through forces acting on material entities, there is no explanation as to what causes those forces or why they exist: masses attract masses with intensity that weakens with the inverse square of the distance between them; electric charges attract or repel each other with similar inverse-square behavior. This formulation allows physicists to describe what masses and charges do in a huge variety of situations. But we do not know *what* electric charge or mass is, or why some fundamental material entities, such as an electron or a quark, have both charge and mass. These are attributes to the material entities we discover with our experimental tools, labels to classify their different types and physical properties. Mass and charge do not exist per se; they only exist as part of the narrative we humans construct to describe the natural world. The same way that these concepts didn't exist five hundred years ago, they may be superseded by a very different set of concepts five hundred years from now. To put it differently, other intelligences, if they exist out there in the cosmos, will no doubt come up with explanations to the phenomena they witness. But to imagine that they will use the same concepts we do is to be hopelessly narrow-minded and anthropocentric, thinking that there is some kind of universal truth to the descriptions we invent.

The understanding of matter and its interactions would radically change in the twentieth century with the widespread introduction of "fields" as a new explanatory device—a new ontology. Particles of matter become localized fluctuations of the fundamental fields, lumps of energy that pop in and out of existence from a preexisting field primer. And even as our understanding of matter and its interactions did improve remarkably once fields became the explanatory instruments of a more fundamental physical reality, we still must interpret them as just another level of description and not as a definitive explanation as to why masses and charges do what they do. All we can say is that at the level of our current understanding, masses and charges are measurable properties of the particle-like excitations of the underlying fields. The success of the present explanation does not preclude the appearance of a better one in the future. In fact, given the way scientific knowledge advances, we have every reason to suppose that this will indeed be the case: just as the electron of one hundred years ago is different from the electron of today, today's electron will be different from the one a hundred years in the future.[3]

Let's go back to the nineteenth century, two hundred years after Newton's science had shaken up the foundations of human knowledge and had redefined the world. It owes such a feat to the remarkable creativity of scientists and their equally remarkable diligence and experimental prowess. In 1865 James Clerk Maxwell unified dozens of apparently disconnected electric and magnetic phenomena as different manifestations of an undulating electromagnetic field. In 1886 Heinrich Hertz confirmed Maxwell's predictions that such electromagnetic undulations did indeed propagate in space carrying energy and momentum. Subsequently, Hertz showed that these electromagnetic waves traveled at light's speed, also as Maxwell had predicted. Theory and experiment formed a rock-solid, formidable relationship. So as to shake off any vestiges of the old "philosophy," natural philosophy became what we now call "science." The *Oxford English Dictionary* dates the word "scientist" to 1834.

A scientist searches for knowledge of the natural world through a specific methodology: the scientific method of hypotheses and experimental validation. Scientists have a clear goal: to describe natural phenomena with rational explanations based on repeatable experimentation and consensus. Speculation is allowed only insofar as it leads to verifiable predictions. A sharp boundary was thus erected between the old natural philosophy and the new science;

those who dare cross it do so at their own peril. Indeed, few have done so. The vast majority of research physicists focus on exploring the properties of solid matter, of matter's most elementary constituents, of fluids and plasmas, and of celestial objects, from planets and stars to galaxies and their distribution in space. However, as our understanding of the cosmos has advanced during the twentieth century and into the twenty-first, scientists—at least those with cosmological and foundational interests—have been forced to confront questions of metaphysical importance that threaten to compromise the well-fortified wall between science and philosophy. Unfortunately, this crossover has been done for the most part with carelessness and conceptual impunity, leading to much confusion and misapprehension. When well-known cosmologists pronounce that "philosophy is useless" or that "quantum cosmology proves that God isn't necessary," they only make matters worse. To understand how this unfortunate state of affairs came about and explore how it further illuminates the limits of knowledge we must first take a short tour through modern cosmology, from the Big Bang to the multiverse.

CHAPTER 8

THE PLASTICITY OF SPACE

*(Wherein we explore Einstein's special and general theories
of relativity and their implication for our understanding
of space and time)*

On November 7, 1919, Londoners woke up to dramatic head-lines from the *Times*: "Revolution in Science. New Theory of the Universe. Newtonian Ideas Overthrown." Likewise, the *New York Times* reported three days later: "Lights All Askew in the Heavens; Men of Science More or Less Agog over Results of Eclipse Observations. Einstein Theory Triumphs. Stars Not Where They Seemed or Were Calculated to Be, but Nobody Need Worry." The news catapulted Einstein to stardom, as they related how two teams of astronomers confirmed his remarkable *general* theory of relativity observing a total solar eclipse in the western coast of Africa and in the town of Sobral, in northeastern Brazil.

Einstein's theory offered a new way to think about the nature of gravity. Instead of the mysterious action-at-a-distance of Newton's theory, Einstein suggested that gravity, or its universal attractive effect, is due to the curvature of space around a massive body. Space becomes elastic, deformable in proportion to how much mass is in a certain region: small-mass bodies create small deformations around them, while massive ones can do more. Thus, while around a human body the deformation is imperceptible (although existing), around the Sun it is much more pronounced. The eclipse test measured light from distant stars as they passed near the Sun. The targeted stars were those with light traveling on a path that grazed the Sun on its way to Earth. The eclipse was a clever way to block the sunlight so that astronomers could see distant stars and measure their positions, comparing the results to similar measurements performed when the Sun was not in between. If space around the Sun was deformed, these paths would be bent, and the stars would appear to be at a different position. Einstein used his theory to calculate the apparent difference in the position of the stars as a result of the Sun's presence. The results from the eclipse measurements were not crystal clear but good enough to prove his theory.

The equations from the general theory of relativity can be used to compute the deformation of space around any massive body, not just the Sun. As light from a distant source travels toward us, it bends this way and that, responding to the bumps and humps of the intervening space.

In a related test, Einstein used the curvature of space caused by the Sun to explain well-known anomalies in the orbit of Mercury that Newton's universal gravitation could not. The theory's success was sealed; it was quickly deemed one of the greatest intellectual achievements in history.

Not just space but also time responds to the presence of matter. In his *special* theory of relativity of 1905, completed ten years before the general version, Einstein had shown how space and time should not be viewed as absolute quantities, as they had been since Newton. They also shouldn't be viewed as separate quantities but as forming a whole, a "spacetime continuum" as it is called, in which time is considered a fourth dimension. Thus, the presence of matter (or energy, in general) deforms space and time, or better, spacetime.

The idea of a spacetime continuum is simpler than it seems. Imagine that you see a fly and, five seconds later, kill it. When you saw the fly, it occupied a point in space, and your "fly clock" marked zero seconds. When you killed it, it was at a different point in space, and five seconds had passed. To locate where and when the fly met its demise you need a position in space and a moment in time. To make the time dimension into one with units of distance, multiply time by a speed. Einstein chose the speed of light, which he assumed was the fastest speed in Nature. In empty space, the speed of light is 186,282 miles per second and is usually represented by the letter c. (Why c? Because *celeritas* in Latin means "speed." We also use the same source word in "acceleration.") In the time you blink your eyes, a beam of light goes seven and half times around the Earth. If you multiply a measure of time, represented by the letter t, by the speed of light, you get ct, which has units of distance. A point in four-dimensional spacetime has coordinates (ct, x, y, z), where x, y, and z are for the locations of a point in three-dimensional space (north-south, east-west, up-down). A sequence of points in spacetime tells a story, say the five seconds between when you saw the fly and where it went until you killed it. This story, or path, in spacetime is called a "world line."

To build his argument, Einstein cleverly focused the discussion on the observer, that is, the person (or instrument) making measurements of distance and of time intervals. He maintained that when two observers were in motion with respect to one another, they would disagree on their measurements of distances and of time intervals. For his special theory, Einstein concentrated on relative motions with constant speeds only. (The general theory would include accelerated motions.) The theory offered a way for the two observers to reconcile their differences. The disagreements are usually minuscule, being determined by the ratio of the relative speed between the observers (v) and the speed of light (c), that is, by v/c. Only for motions with speeds approaching the speed of light are differences noticeable. Nevertheless, they are there, representing yet another layer of deception in our perception of the world. Moving objects appear shorter in the direction of their motion, and moving clocks tick more slowly. For example, an object moving with

60 percent of the speed of light would appear 20 percent shorter. A clock moving with the same speed would slow down by the same amount. As the relative speed between the two observers increases, the shrinking and time delay become more extreme until, as the speed of light is reached, time would come to a halt, and the object would shrink to nothing.

This bizarre situation, however, never happens, since there is yet another effect from relative motion, that of the increase of mass with speed: the mass of an object in motion increases to an infinite value as the object reaches the speed of light. Since to accelerate an object with growing mass requires progressively more energy and is pretty much impossible as the mass gets close to infinite, Einstein's special theory of relativity tells us that no object with mass can ever reach the speed of light; only something without mass, such as light itself, can do so. Furthermore, and completely unreasonably, light always travels at the same speed in the same medium—say, vacuum, air, or water—relative to any observer moving with any speed (under c). To a batter, a baseball slows down when pitched against the wind and speeds up when pitched with the wind. If the pitcher is running in the direction of the batter when he throws the ball, the ball will gain speed, since speeds increase or decrease when compounded. Light's speed, however, is completely indifferent to how its source is moving: its speed is an absolute of Nature, an unchanging constant. In reality, the theory of relativity is a theory of absolutes, of things that don't change in Nature: the laws of physics and the speed of light.[1]

The special theory of relativity lets different observers account for the way Nature operates assuming that the speed of light is always constant and that it is the fastest speed with which signals (and thus information) are communicated. While in Newton's theory space and time were absolute and any speed was possible, when Einstein assumed that light was the absolute speed limit in Nature, he did away with absolute space and time. Going back to Plato's Allegory of the Cave, Newton's theory would be a projection on the cave's wall, an illusion that creatures ignorant of the effects of the constant speed of light would see, believing it to be the correct description of reality. This, of course, is the cave we live in, given that we are myopic to the corrections from light's speed. However, the view from the special theory of relativity is yet another projection on the cave's wall, corrected when *accelerated* motions between observers were taken into account, as they were in the general theory of relativity. After Einstein came up with his general theory, worldviews were again adjusted, as our description of reality inched toward the light: the cave has many walls—no one knows how many—one inside another like Russian nested dolls. As we move from wall to wall, it becomes clear that new layers of description will keep appearing as more is learned about physical reality. All we see are projections on the cave wall. Plato dreamt of a cave with an exit to the light of perfect knowledge, but it seems wise to accept that no knowledge can be perfect or final.

How can something without mass exist? Light is perhaps the greatest mystery. Einstein, one of the key players in helping us understand the physical nature of light, frequently confessed his amazement at its bewildering properties. We don't know why light travels as waves in empty space, while other waves such as sound or water waves need a material medium to support them. We don't know why light's speed has the value that it does or why it is the fastest speed in Nature. All we can say is that so far we have no reason to think of light differently. Once light's properties are taken into account, all sorts of bizarre predictions are unleashed: shrinking distances, slowing time, growing masses. . . . Remarkably, they have all been confirmed in countless experiments. The GPS in your jogging watch or your car owes its precision to corrections to Newton's theory based on both the special and the general theories of relativity. They changed the way we think about space, time, and matter, and they changed the way we think about the Universe. Again, it was Einstein who took the first step.

CHAPTER 9

THE RESTLESS UNIVERSE

(Wherein we explore the expansion of the Universe and the
appearance of a singularity at the origin of time)

"If space is plastic," Einstein reasoned, "and if its plasticity responds to the amount of matter in it, if I knew how much matter there is in the whole of space and how it is distributed, I could use my equations to compute the shape of the Universe." As we noted before, Einstein took a huge step when, just one year after publishing his new theory of general relativity, he applied it to the entire cosmos. As Newton had done with his universal theory of gravitation, Einstein extrapolated his new theory of gravity from the solar system—where it was tested—to the Universe, confident that the same physical principles applied everywhere. He supposed space to be static and spherical and proceeded to simplify things the best way he could. Since it is impossible to know the detailed distribution of all the matter in space, Einstein assumed, very reasonably, that, on average and at large enough volumes, the matter was distributed the same way everywhere.[1] The approximation only works for truly enormous volumes, encompassing many, many galaxies and extending for millions of light-years. Mathematically, it means that the density of matter, the amount of matter in a volume, is approximately constant: bigger volumes imply more matter in the same proportion. That being the case, and since Einstein's

67

equations determined the geometry of space based on the distribution of matter, the geometry had to respond to this homogeneous arrangement, being the simplest shape possible: a sphere. Einstein managed to compute the "radius" of his spherical cosmos and, to make his model stable, added that strange constant we now call the "cosmological constant." He then left things at that, content that his new theory, albeit with a few adjustments and assumptions, could address one of the oldest questions ever asked, What is the shape of the cosmos?

In 1929, only twelve years after the publication of this pioneering paper of modern cosmology, everything changed. The American astronomer Edwin Hubble published his observations of distant galaxies, showing that they were receding from our Milky Way with velocities proportional to their distances: a galaxy twice as far moved away twice as fast. Hubble had at his disposal the largest telescope at the time, the hundred-inch reflector at Mount Wilson in California.[2] With it he was able to see farther and more clearly than anyone else. About a decade earlier, Vesto Slipher had produced evidence that light from distant galaxies seemed to be shifted toward the red when compared to closer ones, an effect known as "redshift." But what could this mean? The answer had been found in the nineteenth century by the Austrian physicist Christian Johann Doppler. Every wave gets stretched if its source (or, equivalently, the observer) is moving away. We know this from our experience with sound waves, as when an ambulance zooms by and we hear its blaring siren going down in pitch. Conversely, if the ambulance is approaching, the pitch goes up. Doppler had proposed this effect in 1842, and musicians blowing their horns on a steam train helped to confirm it in 1845.[3] The same "Doppler effect" happens for light waves, but now instead of pitch we use frequency: the difference between blue and red light is only one of frequency, blue having a higher frequency than red. So when astronomers talk of redshift, they mean the stretching of light waves as a result of the receding motion of the source. A blueshift would imply the opposite, that the source (or the observer) is approaching. Thanks to Doppler, an improbable association is thus established between the mundane and the grandiose: every time you hear an ambulance zooming away on the road, think of the billions of galaxies zooming away in the sky.

Once again, a powerful new instrument triggered a revolution in our understanding of the cosmos. Even before Hubble, a few theorists had speculated that perhaps the Universe didn't have to be static, that it could change in time. The first to do so was the Dutch Willem de Sitter, who criticized Einstein's seemingly ad hoc assumption of a static cosmos: "All extrapolation is uncertain. . . . We only have a snapshot of the world, and we cannot and must not conclude . . . that everything will always remain as at that instant when the picture was taken."[4] In an effort to understand the behavior of matter in an infinite universe, in 1917 de Sitter proposed a rival model in which he considered space to be mostly empty of matter; the only contribution to the geometry of spacetime came from the cosmological term Einstein had invented. Solving the equations, de Sitter showed that any material object would move with growing acceleration. A few years later, Alexander Friedmann, a Russian meteorologist who fell in love with Einstein's theory, showed mathematically that there was nothing in the equations of general relativity forcing the Universe to be static: it could be growing or shrinking in time, like a party balloon. In that case, the density of matter would also change in time, decreasing with the expansion (like when moving furniture from a small to a large room and getting all that extra space) and growing with the contraction. Hubble's discovery of the linear expansion law (that receding speeds of distant galaxies are proportional to distances) showed that Friedmann was right: there was no need to impose a static cosmos or, for that matter, the unnatural constant that kept it so.[5]

An expanding Universe is a source of much confusion. The naïve (and wrong) image most people have is that of a bomb exploding, with galaxies as shrapnel tossed outwards to the ends of space. Why is this wrong? This image presupposes that space stays fixed and that galaxies are moving on it, when in reality the opposite happens: space itself is expanding, and the galaxies are being carried along like corks floating on a river. It is no coincidence that this cosmic motion is known as "Hubble flow." Sure enough, local gravitational attractions between galaxies and groups of galaxies (aka galaxy clusters) may cause deviations from the flow called "peculiar motions." For example, our huge galactic neighbor, Andromeda, is on a collision course with the Milky Way. Simulations and data from the Hubble Space Telescope time it to about four billion years from now.[6]

Hubble's discovery and its confirmation launched the plasticity of space to new heights. From local deviations around stars, we see that Einstein's theory correctly predicts that space as a whole—at least the entire observable Universe, which is all that we can talk about with certainty—stretches in response to the matter within its confines. Things get interesting when we consider what happens as we reverse the expansion, that is, when we look backwards in time: if space is growing now, in the past galaxies were closer together. The further back in time, the closer galaxies were to one another, until we reach a point where they were, at least in principle, squeezed within the same point in space. But how can that be? How can all that exists fit in a point in space? Perplexity grows when we realize that points, being a mathematical idealization, don't exist. So what's going on? Hubble's expansion predicts a cosmos with a history that started at a moment in the past when nothing makes sense. This point is known as a "singularity."

In the 1960s physicists Stephen Hawking and Roger Penrose showed that, given reasonable assumptions about the properties of matter, any expanding universe must have had a singularity in its past. Here is the quandary: since going back in time implies a smaller volume of space, when matter is squeezed to tinier volumes, its density grows without bound. Just imagine what would happen if a crowded subway car would shrink to the size of a sardine can, and then to the size of a pea, and then to the size of an atom, and so forth. Clearly, the density of matter would become infinitely large, and space, being warped by matter, would become infinitely curved. Time would come to a halt since the singularity is reached at t = 0, the "beginning." No decent theory of physics should produce infinities with impunity. Something must be very wrong.

When mathematicians find a singularity (for example, when you divide any number by zero), they explore its neighborhood to see if there is a way out. For example, don't divide the number by zero, but instead try a very small number. Or perhaps there is a path that avoids the singularity and still gets you to where you want to go, like when you avoid a huge pothole as you're driving down a road. In physics, a singularity usually indicates something quite serious, that the theory you are using can't do its job. Something is amiss, and this something usually involves new physics. For example, using Newton's theory to describe how objects behave when moving close to the speed of lights leads to errors, to false illusions projected on the cave wall. We now know that we must use Einstein's theory of special relativity to get the correct answers. Same with strong gravity: Newton's theory is a great approximation for fairly weak gravitational attraction but needs to be corrected near very massive objects such as the Sun.

No theory is ever complete or final, as new extremes ask for new formulations, and new formulations require experimental validation that, in turn, depends on available technologies. When scientists are hunting for a predicted effect to test a theory, they often would rather find something unexpected that will force them back to the drawing board and possibly to new knowledge. Case in point: most physicists involved in the search for the Higgs boson at the Large Hadron Collider in Switzerland would rather have found a particle somewhat different from the one predicted in the Standard Model of particle physics. The unforeseen leads to change.

The cosmic singularity points to the need for a new physics, beyond what Einstein's general theory of relativity can provide. And

since near the beginning of time distances were tiny, the new physics needs to explain how space, time, and matter behave at very
short distances: the physics of the very large meets the physics of
the very small. This is the realm of "quantum gravity," the marriage
of the general theory of relativity with quantum physics, the physics of atoms and their subatomic components. In an amazing twist,
the study of the Universe and its history led us to an investigation
of the smallest constituents of matter. As we currently see it, the
two are deeply linked: physicists will not understand the origin of
the Universe before they can describe how quantum physics influences the geometry of spacetime. But before we get to that, we must
explore some of the fundamental consequences of modern cosmology to the limits of knowledge. It all starts with the finiteness of the
speed of light and the notion of "nowness."

CHAPTER 10

THERE IS NO NOW

(Wherein we argue that the notion of "now"
is a cognitive fabrication)

What goes on when you see something, say, this book you are reading? Leaving aside the whole business of how the brain processes visual information, let's just focus on the information travel time. To make life simple, let's also just consider the classical propagation of light, ignoring for now how atoms absorb and reemit light. Light is bouncing around the room because either the window is open or the lamp is on, or both. This bouncing light hits the surface of the book, and some of it is absorbed, while some is reflected outwards in different directions. The page and the ink used for printing absorb and emit light in different ways, and these differences are encoded in the reflected light. A fraction of this reflected light then travels from the book to your eyes, and thanks to the brain's wondrous ability to decode sensorial information, you see the words on the book's page.

It all looks instantaneous to you. You say, "I'm reading this word now." In reality, you aren't. Since light travels at a finite speed, it takes time for it to bounce from the book to your eye. When you see a word, you are seeing it as it looked some time in the past. To be precise, if you are holding the book at one foot from your eye, the light travel time from the book to your eye is about one nanosecond, or one billionth of a second.[1] The same with every object you see or person you talk to. Take a look around. You may think that you are seeing all these objects at once, or "now," even if they are at different distances from you. But you really aren't, as light bouncing from each one of them will take a different time to catch your eye. The brain integrates the different sources of visual information, and since the differences in arrival time are much smaller than what your eyes can discern and your brain process, you don't see a difference. The "present"—the sum total of the sensorial input we say is happening "now"—is nothing but a convincing illusion.

Even if nerve impulses propagate fast along nerve fibers, their traveling times are still much slower than the speed of light. Although there are variations for different types of nerves and for different people, the speed is around 60 feet per second. That is, nerve impulses travel about 1 foot in sixteen milliseconds. (A millisecond is one thousandth of a second.) For comparison, light travels 2,980 miles in the same amount of time, a little more than the driving distance from New York to San Diego.

Here is an imaginary experiment that illustrates the implication of these time differences. Imagine two lights programmed to flash simultaneously every second. One of the lights is fixed at 10 yards from an observer, and the other can be moved away on a straight rail. Imagine separating them by increasing distances as they flash together every second. An observer will start perceiving a difference in the flashing times when the distance between the two lights is larger than about 2,980 miles. Since we can't see this far, our perception of the simultaneous now seems very credible for huge separations. An alternative, and more realistic, experiment could be set up to test this theory: have two lights flashing at slightly different times, and check when observers notice a difference. If my conjecture is correct, observers will start to notice differences when the timing interval is larger than about twenty milliseconds or so. This timescale sets the limit of visual simultaneity in humans.

The arguments above lead to a startling conclusion: the present exists because our brain blurs reality. To put it another way, a hypothetical brain endowed with ultrafast visual perception would catch the difference between the two flashing lights much earlier. For this brain, "now" would be a much narrower experience, distinctive from the human "now." So in addition to Einstein's relativity of simultaneity involving two or more moving observers, there is also a relativity of simultaneity at the cognitive level resulting from the subjective perception of simultaneity or "now" for the individual or, more generally, for every kind of brain or apparatus capable of detecting light.[2]

Each one of us is an island of perception. Just as when we look out into the ocean and call the line where water and sky meet the *horizon*—a limit to how far we see—our perceptual horizons comprise all the phenomena that our brains compute as happening simultaneously even if they are not: the perceptual horizon delineates the boundary of our "sphere of now." Since light is the fastest speed in Nature, that's the one I'm using to define our sphere of now. (Had we used the speed of sound, of only 1,126 feet per second in dry air and at 68 degrees Fahrenheit, the sphere of now would have a much smaller radius. Two lightning strikes miles apart look simultaneous but wouldn't sound simultaneous.)

To summarize: given that the speed of light is fast but finite, information from any object takes time to hit us, even if the time is tiny. We never see something as it is "now." However, the brain takes time to process information and can't distinguish (or time-order) two events that happen sufficiently close to one another. The fact that we see many things happening now is an illusion, a blurring of time perception. Since no brain is the same, every person will have their own limits of time perception and their own sphere of now. In fact, every brain, be it biological or mechanical (light-sensitive detecting device), has a different processing time and will have its own sphere of now; each one will have a distinctive perception of reality. From current neurocognitive experiments, it seems reasonable to suppose that on average a human's time perception is on the order of tens of milliseconds. The distance light travels in this time interval is the approximate radius of an individual's sphere of now—a few thousand miles.

"Now" is not only a cognitive illusion but also a mathematical trick, related to how we define space and time quantitatively. One way of seeing this is to recognize that the notion of "present," as sandwiched between past and future, is simply a useful hoax. After all, if the present is a moment in time without duration, it can't exist. What does exist is the recent memory of the immediate past and the expectation of the near future. We link past and future through the conceptual notion of a present, of "now." But all that we have is the accumulated memory of the past—stored in biological or various recording devices—and the expectation of the future.

The notion of time is related to change, and the passage of time is simply a tool to track change. When we see something moving in space, we can follow how its position changes in time. Say it's a ball; as the ball moves, it will describe a curve in space, an imagined sequence of points from initial position A to final position B. We can tell where the ball is between A and B by ordering its location sequentially in time: at zero it is leaving the soccer player's foot—point A; at one second it is hitting the upper-left-hand corner of the goal—point B. The curve in between A and B links the position of the ball at the intermediate times between zero and one second. A ball, however, never occupies a single point in space, and time can never be measured with infinite precision. (The most accurate clocks use electronic transitions in atoms to achieve a precision of about one billionth of a second per day.) Mathematically, though, we brush all this aside and compute how the position of the ball changes in time *instantaneously*: for every moment of time we claim to know its position. Clearly, this is only an approximation, albeit a very good one.

We represent the flow of time continuously so that each instant of time has a (real) number attached to it. In our example of the soccer ball, time will cover the number line from zero to one. How many instants of time are there between zero and one second? Mathematically, there is an infinite number of them, since there are infinite numbers between zero and one. (You can keep subdividing intervals into smaller and smaller bits: a tenth of a second, a hundredth of a second, a thousandth of a second, and so on.) But even the most accurate clocks have limited precision. We may represent time continuously, but we measure it in discrete chunks. As a consequence, the notion of "now," a time interval of zero duration, is only a mathematical convenience having nothing to do with the reality of how we measure time, let alone perceive it. I will have more to say about this and what it means about our notion of reality when we get to quantum physics, where nothing is ever continuous.

CHAPTER 11

COSMIC BLINDNESS

*(Wherein we explore the concept of cosmic horizons and
how it limits what we can know of the Universe)*

As we move on to modern cosmology, things get even more interesting. The combination of having a Universe with a finite age—the time elapsed since the Big Bang—and the finite speed of light creates an insurmountable barrier to how much we can know of the cosmos. This is a new and different kind of limitation from the ones we have discussed so far, as it does not depend on the precision of our measuring tools, on our "myopic" view of reality. This is an absolute limitation on how much we can know of the physical world, a limitation that Galileo, Kepler, and Newton couldn't even guess existed. The Universe may be spatially infinite, but we can't ever be sure. We live in a spherical bubble of information, like fish in an aquarium. There is a beyond, we can infer its existence through our blurred vision, but we can't know what's "out there." Three centuries ago, de Fontenelle already knew that the agony and the ecstasy of philosophical and scientific inquiry come from wanting to know more than we can see. We stretch out and approach the boundary even at the danger of smashing our heads against the glass. Just as our predecessors did, we want to break free and probe the new unknown. But now we can't. The "out there" has become an unknowable.

Einstein's theories of relativity place some discouraging limitations on traveling backwards in time. The special theory states bluntly that it is impossible to do so, since masses grow without bound as we reach the speed of light. As my then six-year-old son Lucian proudly announced to me the day I wrote these lines in one of our frequent metaphysical bouts while driving to school, "There is only one thing that can travel at the speed of light, Dad: light!" Precisely. And it can do so because it has no mass. Unlike a chunk of matter, which even at rest has energy equal to its mass (m) times the square of the speed of light (c^2), as Einstein expressed in his famous $E = mc^2$ formula, light is never at rest. Its energy is solely fixed by its frequency (f) in a disarmingly simple equation, $E = hf$, where h is the Planck constant, a very tiny constant of Nature that sets the stage for all things quantum: the higher the frequency of the light, the higher its energy. The formula $E = hf$ is not telling us something about the behavior of light as we see it around us, made of bouncing waves. This formula hides a mystery, one of the greatest in modern science.

To come up with this formula for the energy of light, Einstein advanced what in his own judgment was his most revolutionary idea: that light could be interpreted as both a wave, as people predominantly thought during the nineteenth century, and a particle. These little light bullets came to be known as "photons," the particles of light. The formula $E = hf$ is for the energy of a single photon. Beams of light will contain many photons, and their energies will be always in integer multiples of the energy of a single photon, hf. This is also what happens with money. Every financial transaction from cents to billions of dollars is in multiples of a cent. Clearly, very large transactions lose track of their "quantumness," of their link to a single cent. But just as every cent is money, every photon is light.[1]

In practice, light beams may combine photons of many different wavelengths. For example, sunlight consists of all visible colors, from red to violet, each color with its own wavelength and related photon. In our monetary analogy, sunlight would be like a customer that comes into a money exchange booth with many different currencies (one for each color), each currency having its own version of a cent (its photon with energy hf).

Most of the information we gather about the Universe comes to us in some form of electromagnetic radiation. Optical astronomy, the noble tradition that collects photons of visible light with the naked eye or with telescopes, is but one example. Nowadays astronomers look at the skies at wavelengths comprising almost the entire electromagnetic spectrum, from radio to gamma rays. However, for every type of light we look at, the same limiting speed applies.[2] So just as when you look at this book you see it as it was one billionth of a second ago, when you look at the night sky, you see it in the past. You see the Moon as it was 1.282 seconds ago, since its average distance from Earth is 1.282 light-seconds. The Sun you see as it was 8.3 minutes ago, since the Earth-Sun distance is about 8.3 light-minutes. The Sun could explode right *now*, and you wouldn't know it for eight minutes; the last thing you'd find out . . .

Moving further out in the solar system, we experience the complication that the planets orbit the Sun at different speeds: distances between the Earth and the other planets can vary a lot, depending where they are in relation to each other along their orbits. For example, the distance from Earth to Mars varies from about 4.15 light-minutes (point of closest approach, both planets at the same side of Sun) to 20.8 light-minutes (point of farthest distance, planets at opposite side of Sun). Unless you are a NASA engineer projecting a flight to another planet, it's best to refer to distances to the Sun. Mars is about 12 light-minutes from the Sun, while Neptune is about 4.16 light-hours. We see that the 8 light-minutes between Earth and Sun become a small correction as we look at the outskirts of the solar system. The farthest known structure in the solar system is the Oort cloud, a belt of icy balls that frames the Sun and planets at about 1 light-year. There lie the last bits and pieces of the gas cloud that contracted 4.6 billion years ago to form the Sun and its court of planets and moons.

Every celestial object within this 2-light-year diameter bubble, including us, shares a common origin. As we move further out from the Sun we get into alien territory, other stars with their own courts of planets and moons, each with their own shared origin and history. We can think of these star systems as different families, with children that share a common parent (the originating gas cloud) but that go on to do their own thing. The closest star system to the Sun is in the constellation of the Centaur, or Centaurus, that Ptolemy had already listed in the second century CE. This means that you can see it with the naked eye in the southern sky and try to convince yourself that indeed it looks like a creature that is half man and half horse. In Centaurus we find the closest stars to the Sun, a triplet of stars collectively named Alpha Centauri, at a distance of about 4.4 light-years, some twenty-six trillion miles. Of the three, the closest is Proxima, at 4.24 light-years from the Sun. So when you look at Alpha Centauri (and mistakenly think it's a single star in the sky), you are receiving information from what the star trio looked like some 4.4 years ago. For all we can tell, the stars may not even be there anymore. We can infer that they still are because we know what kinds of stars they are and how far down they have evolved along their life cycle. But we don't have—and can't have—direct proof that they are still there. The night that we see now is a collective of past histories.

For people from the Southern Hemisphere, Centaurus borders the famous Southern Cross (or Crux) on three sides. I was born in Brazil, so no celestial marker is more meaningful to me. (Orion is a distant second.) The Cross figures prominently in our flag (and that of Australia, New Zealand, Papua New Guinea, and Samoa), representing our allegiance to the sky, a reverence for our celestial roots. It no doubt nourished the faith of the pious and greedy Catholic explorers that arrived in South America in droves during the early sixteenth century, convinced that the cross in the sky was a sign from God, that this was a promised land filled with riches and beauty. So advised, they went ahead and sacked it the best way they could.

Drawing an imaginary line linking the two vertical stars of the Cross and extending it down leads to a point close to the Southern Celestial Pole. Every time I visit Brazil, even after living for so long in northern latitudes, I look up and search for it. Only then do I feel that I have truly arrived at the part of the sky I called home for so long. Funny to think that the stars making up this constellation are at various distances averaging hundreds of light-years from us; what we see as a cross is really an illusion, projected on the imaginary celestial dome.

For those who are enthusiasts of aliens and space travel, here is a sobering thought. If we used our fastest spaceship in an expedition to Alpha Centauri, rounding up its speed to about thirty thousand miles per hour, it would take roughly one hundred thousand years for it to get to Alpha Centauri. Even if we developed some new technology capable of taking us through space at one-tenth of the speed of light, it would still be forty-four years before we got there. So unless we organize massive transgenerational migratory expeditions taking thousands of years, or develop unanticipated new space travel technologies, we won't be visiting other stellar systems soon. And that's only to our closest celestial neighbors.

Our galaxy, the Milky Way, has a diameter of one hundred thousand light-years: if you turned on a flashlight at one end, it would take that long for photons to get across. In other words, as we peruse stars in the boundaries of our galaxy, we are seeing them as they were when *Homo sapiens sapiens*, our subspecies of "modern human," was claiming its ground on Earth. Jump to Andromeda, our galactic neighbor, and light from there left some two million years ago, when our first *Homo* ancestors were spreading across Africa.

As astronomers look out into space, they are truly looking at the past, collecting light that left its source millions—even billions—of years ago. This picture holds for an expanding universe, although things get a bit more complicated. In a static universe, what you see is what you get: if you know how far an object is, you can compute how long ago light left it simply by dividing its distance by the speed of light. Because the expansion of the Universe carries galaxies and other light sources with it, the light they emit can travel farther in the same amount of time than it would in a static space. Think how a swimmer going downstream in a river gets a ride from the current and goes farther then he would in a swimming pool in the same amount of time. In our expanding Universe, the light of an object that is now, say, 2.6 billion light-years away from us left it 2.4 billion years ago and not 2.6 billion years ago. The discrepancy increases for objects at greater distances. At the time of this writing, the record for the most distant object detected goes to a galaxy 32.1 billion light-years away. Light has left it 13.2 billion years ago, traveling almost 2.5 times farther than it would have in a static universe. Keeping in mind that the Universe is about 13.8 billion years old, light from this object traveled to us for most of cosmic history, leaving its source only about 600 million years after the Big Bang.

I'm sure the reader knows where I am going with this. At some point we must hit a brick wall, the glass of the aquarium, a barrier we cannot go beyond. In principle, this point is the singularity, the beginning of time. In practice, the wall, at least from the perspective of collecting information from electromagnetic radiation, comes a bit before we hit the singularity. At about four hundred thousand years after the Big Bang, the Universe underwent a deep transformation. To see why, picture the early Universe as a soup of elementary particles, furiously colliding with one another: photons, protons, electrons, neutrons, and some light atomic nuclei.[3] The earlier in time, the hotter the cosmos, and the more furiously these particles interacted. So moving forward in time, the expanding Universe got cooler: as the Universe grew, particles lost some of their energy. This energy loss and cooling allowed for what was highly improbable earlier on—an electron and a proton joining into a hydrogen atom—to become possible. Before this time, photons from the radiation filling up space were so energetic that whenever electrons and protons tried to get together to make hydrogen atoms, they would hit the electrons and stop the bonding from happening. It was an intense ménage à trois. This cosmic love triangle only got resolved after the photons lost enough of their oomph to allow for the bonding of electrons and protons. The simplest of atoms was finally born; the leftover photons, freed from the difficulties of shared love, could roam unimpeded across space. This process is called "recombination" and marks the transition from an opaque to a transparent universe.[4]

Before recombination, photons were so busy in their love triangle with electrons and protons that they couldn't travel freely across space. And if photons can't propagate across space, we can't detect them. The early Universe was opaque to electromagnetic radiation of any type: trying to see what was going on before recombination is like trying to see through a dense fog. Shortly after recombination, however, these photons finally became free to travel across space. The term used is "decoupling" (of matter and radiation). These decoupled photons, raging across the cosmos, constitute what is known as the "cosmic microwave background radiation," the leftover, cooled-down glow from the epoch when the first atoms appeared. At the time of recombination, the temperature of the radiation was about 4,000 degrees Kelvin, or 7,200 degrees Fahrenheit; the Universe was about as bright as a tubular fluorescent lamp. Let there be light, indeed! After 13.8 billion years of cosmic expansion, these leftover photons cooled down to a horribly frigid 2.75 degrees Kelvin, or −454.7 degrees Fahrenheit. The cosmos has lost its baby hue, and deep space is now immersed in cold blackness.

We can now see how the concept of a horizon makes an entry in cosmology. At the seashore, the horizon marks the limit of what we can see; we know the ocean continues beyond the horizon, even if we can't see it. The same with the Universe. There is a horizon, the farthest point from which light could have reached us in 13.8 billion years, the age of the Universe. Even if space extends beyond it, we cannot receive signals from behind this wall. Relativistic cosmology has forced us to confront a new limitation on how much we can know of the cosmos. The physical Universe mirrors the Island of Knowledge.

The prospect of traveling faster than the speed of light via conventional spaceships is very remote. We have no indication whatsoever that the predictions from the special theory of relativity are wrong. On the other hand, as I am arguing here, we can never know for certain; there is always the possibility that our current construction of causality and time-ordering based on the speed of light is not the final word on the topic. We should build solid arguments based on current scientific knowledge but keep our minds open for surprises. To believe that scientific arguments are invulnerable to change is a trap we should not fall into. As it should be clear from our review of the shifting views on the nature of the heavens, no scientific construction is inviolable. Change is the only way forward.

Everything that we know (and can know) about the Universe comes from information within our cosmic bubble, the causal domain delimited by the speed of light and by our expanding Universe's history. In a somewhat ironic twist, we are, after all, enclosed in a cosmic dome—not necessarily within a finite space, as Aristotle or Copernicus or Einstein would have proposed, but within a temporal one. We are effectively blind to what is "out there," beyond the cosmic horizon, unless things there could somehow send signals in here. Crazy stuff could be happening out yonder, pink droid elephants dancing the samba on planet Mamba, and we wouldn't—couldn't—know about it.

Our current most powerful probe of the early Universe is the cosmic microwave background, the leftover photons from recombination. Years of sensational observations coupling data from satellite missions such as the Cosmic Microwave Background Explorer, the Wilkinson Microwave Anisotropy Probe, and now the Planck satellite, with data gathered in dozens of ground-based efforts, has helped astronomers build a comprehensive map of the infant cosmos. That some of the measurements from the microwave background were independently confirmed by different telescopic surveys shows that modern cosmology has become a serious data-driven science, far removed from its early speculative days. The gravitational pushing and shoving that matter underwent during the cosmic infancy, duly recorded in tiny temperature fluctuations of microwave background photons, reflect, to an amazing degree, the way in which galaxies are distributed in the sky today.

And what do current measurements of our cosmos tell us? First, that the cosmic geometry is flat, like a three-dimensional version of a tabletop (which is two-dimensional): unless light passes close to a massive star or galaxy, it travels in straight lines in any direction it moves. A flat geometry is one of three possible options. The other two are a *closed* geometry, like the surface of a sphere, where a path in the same direction brings you back to your starting point (don't try to see this in three dimensions); and an *open* geometry, which we can try to imagine (poorly) from a two-dimensional analogue like the surface of a Pringles potato chip, curving in two different directions. (A horse saddle is also often used as an example, curving down under the legs of the rider but up along the horse's back.)

The kinds of stuff that exist in the Universe and their relative amounts determine the cosmic geometry, the shape of space at the grandest of scales. Two competing trends fight for control: a tendency to expand—from initially having hot matter and radiation squeezed into small volumes—and a tendency to contract, from the gravitational attraction of all the stuff in it. The winning trend determines the fate of the Universe: the expansion may go on forever, or if there is enough matter, it may reverse and become a contraction. Big Bang reverses into Big Crunch.

Since Einstein has taught us that matter influences the geometry of space, these two trends will determine the cosmic geometry. An underdense universe, one where the attractive pull of all the stuff in it is not strong enough, will expand forever and have an open geometry. The critical amount of energy per volume to halt the expansion is sometimes called the "critical density," only about 5 atoms of hydrogen per cubic meter of space. Measurements indicate that the contribution to the density from normal atomic matter comes to only about 0.2 atoms per cubic meter, well below the critical value (4.8 percent of it, to be precise.)[5]

To normal atomic matter we must add another kind of matter, of still mysterious composition; we call it "dark matter." Why "dark"? Because this kind of matter doesn't shine; that is, it doesn't emit electromagnetic radiation of any sort. We know it exists because we can see how it makes galaxies rotate faster than they would. Astronomers also measure how dark matter, which collects around galaxies like a ghostly veil, distorts space around these galaxies. This distortion of space is quite spectacular. To see it, astronomers look at light coming from very distant objects as it passes close to a nearby galaxy. Just as light from distant stars gets bent as it passes by the Sun, intervening galaxies bend light from these distant objects. The effect is known as "gravitational lensing," since it acts pretty much the way an ordinary lens does, bending light's path as it goes through it. Putting these observations together, including information from the microwave background, the amount of dark matter in the Universe comes to a little under six times that of ordinary matter, contributing to the cosmic density at about 25.9 percent of the critical value. The nature of this dark matter, that is, its composition, remains one of the central mysteries of modern cosmology and particle physics. It is one, though, that we hope we will understand as our measuring tools improve, being thus quite different from the existence of a cosmic horizon, a limit we cannot overcome.

The current leading candidates for dark matter are particles predicted to exist from supersymmetric theories, extensions of current particle physics that include a new symmetry of Nature. The reader may recognize the "super" in supersymmetry from superstring theory, a candidate theory for unifying general relativity and quantum mechanics. As of the winter of 2014, no evidence for supersymmetry had been found, despite decades of intense search and the enthusiastic support of many physicists. At this point, it is unclear and somewhat doubtful that supersymmetry is realized in Nature.

Another explanation for dark matter is to fault Einstein's general relativity, instead of proposing a new kind of particle. The theory imposes a change in the behavior of gravity effective only at large, galactic distances. Here too there is no evidence that such an explanation would work and be consistent with several astrophysical observations. The puzzling nature of dark matter is a powerful illustration of a tool-driven discovery—the existence of a new contribution to the material composition of the Universe—that remains unresolved because of the limited precision and reach of our current tools. We know there is something cloaking the galaxies, but we don't know what it is.

If we only considered the total mass (and energy) from atomic and dark matter and radiation (radiation contributes almost nothing), the Universe would have an open geometry, with only about 30 percent of its critical density. But that's not the whole story. If there is a cosmological constant or something similar in the Universe, its effect would be to make space stretch out. Recall that Einstein came up with this constant to make his closed universe static, and then discarded it after learning about Hubble's discovery of the cosmic expansion law. In a remarkable turn of events, measurements from two groups of astronomers indicate quite strongly that something akin to a cosmological constant not only exists but dominates the stuff inside our cosmic horizon. The measurements were announced in 1998 and shocked the physics and astronomy communities. No one wanted to believe it. But time has passed, and the results have stood countless tests and criticisms. Again, powerful new tools uncovered something we didn't know existed, making the cosmos even stranger. As with dark matter, we know something is out there, but we don't know what it is.

In 2011, three of the groups' leaders received the Nobel Prize in Physics for their discovery of "dark energy," a truly mysterious entity that acts like a cosmological constant and is responsible for not just stretching space but doing so at an accelerated rate. More remarkably, once the contribution of dark energy to the density of the Universe is computed, the number comes to a little under 70 percent of the critical density. Adding all different contributions, you find something amazing: not only does dark energy far outweigh everything else in the cosmos, shining or dark, but the total adds to the critical density. It almost sounds too contrived to be true. Of all possible values, the density of stuff in the cosmos is nearly identical to the critical value. Current measurements set the total energy density of the Universe equal to the critical value with an accuracy of about 0.05 percent.

At first sight, a cosmos right at its critical density now does seem like the work of divine tweaking. But more careful thinking reveals that universes capable of generating life must satisfy strong constraints: they can't be too underdense, or they would expand too fast, and matter wouldn't condense gravitationally into stars and galaxies; they can't be too overdense, or they would have collapsed onto themselves long before stars were born. Universes capable of harboring life must reach an old age so that stars go through several generations to produce heavy chemicals with high enough abundance. These constraints limit the possible values that the density of the universe and a hypothetical cosmological constant could have. An optimal universe would be one sitting right at the critical value of matter density, as is our case. Physicist and author Paul Davies called this the "Goldilocks Universe," and indeed it is tempting to consider that our cosmos is "just right" for life. I offer a somewhat different interpretation of this cosmic coincidence, to which I will turn soon.

The current measurements are so precise that the density of matter and dark energy are known with an accuracy of better than one-half of 1 percent. Unless something dramatic happens and the dark energy weakens its cosmic dominance in the future, the data indicates that we live in a flat universe fated to expand forever with ever-increasing acceleration. If the Universe continues with its accelerated expansion, a bleak future awaits our (very) distant descendants. As space keeps on stretching, it will carry away most celestial luminaries, the galaxies that are now within telescopic sight. With time, their receding speeds will surpass the speed of light, and a new cosmological horizon will appear, beyond which their light will no longer reach us.[6] Eventually, only our Local Supercluster, the large agglomerate of galaxies including the Milky Way and Andromeda that are bound together by their mutual gravitational attraction, will be visible in the night, albeit much transformed from its current shape. As noted, in a few billion years, the Milky Way and Andromeda galaxies may merge and become one large galaxy. In about four billion years, our Sun will turn into a red giant, and life on Earth will be impossible. (Actually, this will happen much before then, because of instabilities in the solar energy output.) If cosmologists from this very faraway future don't have access to past observations, they will conclude that they live in a Universe quite different from ours: with no receding galaxies in sight, they won't ever know of the expanding Universe, or of the Big Bang. Ironically, their cosmology will return to that of a static cosmos, an island universe comprising the galaxies within the Local Supercluster, surrounded by a vast expanse of dark, empty space. Their Island of Knowledge will shrink and eventually disappear altogether, mirroring the growing darkness surrounding it. In time, the few stars still able to produce light will grow old and fade away from sight. The cosmos will turn nearly black, and Lord Byron's nightmarish vision will become a reality:

> I had a dream, which was not all a dream.
> The bright sun was extinguish'd, and the stars
> Did wander darkling in the eternal space,
> Rayless, and pathless, and the icy earth
> Swung blind and blackening in the moonless air;
> Morn came and went—and came, and brought no day,

And men forgot their passions in the dread
Of this desolation; and all hearts
Were chill'd into a selfish prayer for light[7]

Fortunately, current data places this despairing prospect far into the future, possibly one or two trillion years from now. I don't mention it as something for us to worry about but to reflect upon, as it has implications for our current view of the cosmos. The Universe we measure tells only a finite story, based on how much information can get to us (the cosmic horizon placing a limitation here) and on how much of this information we manage to gather (our technological prowess placing a limitation here). Those unfortunate cosmologists of the future, if basing their science only on what they could measure, would construct an erroneous picture of the world, missing the fact that their bleak predicament has an explanation tracing back trillions of years into the past. Their static cosmos would be an illusion, a consequence of their living within a cosmological horizon where galaxies don't follow an expansionary trend, as they do in ours. The lesson here is distressing: not only are there causal and technological limits to how much we can know of the cosmos, but what information we do manage to gather may be tricking us into constructing an entirely false worldview. What we measure doesn't tell us the whole story; in fact, it may be telling us an irrelevantly small part of it.

To avoid the funk of a modern scientific nihilism, we must find joy in what we are able to learn of the world, even if knowing that we can only be certain of very little. Grand statements such as "This is the true nature of the Universe" should be downgraded to "This is what we can infer about the nature of the Universe." The word "true" doesn't have much meaning if we can't ever know what it is. Clearly, what we do infer is still quite spectacular, and that's what should have value. Undeterred, we must push beyond the boundaries of our cosmic horizon and delve into what may lie out there, a multiverse of endless expanse.

CHAPTER 12

SPLITTING INFINITIES

(Wherein we begin to explore the notion of the infinite, and
how it translates into cosmology)

"What's infinity plus infinity?" asked my son Lucian. "Infinity," I replied stoically.

"But how can a number plus itself be itself?" Lucian insisted. "I thought only zero could do that."

"Well," I said, "infinity is not really a number. It's more of an idea."

Lucian rolls his eyes and thinks. "So infinity plus infinity is infinity, even if infinity is the opposite of zero?"

"Yep."

"Weird, Dad."

"Yep."

Infinite is that which is beyond countable, although mathematicians often refer to countable infinity as opposed to uncountable infinity. Yes, there are different kinds of infinity. For example, the set of all integer numbers (. . . ,–3,–2,–1, 0, 1, 2, 3, . . .) is a countably infinite set. Another example is the set of rational numbers, numbers of the form p/q constructed from fractions of integers such as 1/2, 3/4, 7/8, etc. (Excluding division by zero.) The number of objects in each of these sets (also called the "cardinal" of the set) is called "aleph-0." *Aleph* is the first letter of the Hebrew alphabet and has the cabalistic interpretation of connecting heaven and earth: $\aleph$. Aleph-0 is infinite, but not the largest possible infinite. The set of real numbers, which includes the sets of rational and irrational numbers (those numbers that cannot be represented as fractions of integers, such as $\sqrt{2}$, π, e), has a cardinal of aleph-1. Aleph-1, known as the "continuum," is larger than aleph-0, and can be obtained by multiplying aleph-0 by itself aleph-0 times: $\aleph_1 = \aleph_0^{\aleph_0}$). The German mathematician Georg Cantor, the pioneer who invented set theory and developed such concepts, stated the "continuum hypothesis": there is no set with a cardinal between aleph-0 and aleph-1. However, current results imply that the continuum hypothesis is undecidable; that is, it is neither provable nor unprovable. The human mind gets muddled by ideas of different infinities, even within the formal rigidity of abstract mathematics. But we shall return to the issue of undecidability in due course. For now, let us just take the notion of countable and uncountable infinities to the cosmos.

Is space infinite? Does the Universe extend toward infinity in all directions, or does it bend back on itself like the surface of a balloon? Can we ever know the shape of space? The existence of a cosmic horizon and the fact that we only receive information from what is within our light bubble places a serious limit on what we can learn from beyond its edge. When cosmologists state that the Universe is flat, what they really mean (or should mean) is that the portion of the Universe that we measure is flat or very nearly so within the precision of the data. The measured cosmic flatness does imply that the Universe is much bigger than what we can measure. But we cannot, from the flatness of our patch, make any conclusive statements about what lies beyond our information bubble or about the global shape of the Universe.

Of course, we may and should speculate, and possibly learn something, about the "outside" from in here. Going back to the beach may be useful. Unless you believe, like the Mesopotamians, that the horizon is the edge of the world and beyond it you will find only oblivion, by standing at the beach and looking at the horizon you can infer that there is more ocean beyond. When a ship is at the horizon, its lower part is invisible because of the curvature of the Earth. Or you could spot an island in the distance and take note of its position with respect to the horizon. Then, climbing a high mountain, you may observe that there is more ocean beyond the island and thus confirm that the horizon as seen from the beach does not mark the end of the ocean. However, even from a very high mountain no one can figure out our planet's complicated geometry of oceans and continents, or that it is shaped like a sphere slightly flattened at the poles. Historically, the extent to which we could move along (and above) our planet has determined our view of it. The alliance between mathematics and astronomy helped tremendously, as illustrated in Eratosthenes's estimate of the circumference of the Earth around 200 BCE. Also, during a lunar eclipse Earth throws a circular shadow on the Moon. There are many other examples. But the conclusive proof of Earth's sphericity had to wait until Ferdinand Magellan and Sebastian Elcano completed its circumnavigation in 1521. People may (even if wrongly) doubt a mathematical argument or what shapes a faraway shadow, but they can't doubt a closed-loop path. Unfortunately, when it comes to the cosmic horizon, a circumnavigation is out of the question.

A two-dimensional analogy may be useful. Imagine the surface of a very large ball. Imagine further that there are creatures living on a galaxy on this surface. As with our Universe, this two-dimensional universe also had a Big Bang in its past. As we have a cosmic horizon, so do the creatures living on the galaxy. Their horizon would be a disk-shaped patch on the ball's surface. If the ball is huge and the patch relatively small, the creatures would infer that their universe is infinite and its geometry is flat. (To see this, draw a small circle on a large balloon. The enclosed area will appear flat.) But this conclusion, of course, is incorrect. Even if they measure the geometry of their patch to be flat, they live in a finite universe—the surface of a ball. Could these creatures ever find the truth about the shape of their universe if unable to venture beyond their patch?

In the same way, can *we* find the global shape of the Universe even if we are stuck within a flat cosmic horizon? If our Universe is shaped as a three-dimensional sphere, we may be out of luck. If we judge from current data, the curvature of the sphere would be so tiny that it would be hard to measure any indication of it. An interesting, although far-fetched possibility, is that the Universe has a complicated shape, with what geometers call a "nontrivial topology." Topology is the branch of geometry that studies how spaces can be continuously deformed into one another. "Continuously" means without cutting, as when you stretch and bend a rubber sheet. (These transformations are known as "homeomorphisms.") For example, a ball with no holes in it can be deformed into a football-shaped ellipsoid, a cube, or a pear. But it cannot be deformed into a doughnut. A ball with one hole in it can be deformed into any hat, as long as the hat has one hole. A doughnut can be deformed into a mug with a handle. The idea, then, is that a complicated cosmic topology may leave its signature imprinted in things we can measure. For example, if the topology is nonsimply connected (e.g., when there are holes in the surface, as in a doughnut), light from distant objects may produce patterns in the microwave background. Specifically, if the Universe is doughnut-shaped and its radius is small compared to the horizon size, light from distant galaxies may have had time to wrap around a few times, creating multiple identical images like reflections in parallel mirrors. In principle, such ghostly mirror images or patterns could be seen, providing information about the global shape of space.

As an illustration of how the precision of observations allows for healthy speculation, unless we have absolute certainty that the curvature of our cosmic patch is exactly zero, there will always be room for topologies that depart from the trivial, flat, three-dimensional space. It is, of course, possible that ghost mirror images will be detected, and we will then have some ground to consider that the global shape of the Universe is not simply flat. Consider, however, the other, more interesting possibility, that we detect no such images. Could we then conclude that space is flat? Since we can never measure anything with *absolute* precision, even if current data strongly points toward zero spatial curvature within our cosmic horizon, we can never be certain. In the absence of a positive detection of curvature, the question of the shape of space is thus unanswerable in practice. Is it an unknowable? It seems to be, unless something quite drastic comes to the rescue. If the Universe were shaped like a sphere, as Einstein wanted, and if it would collapse on itself in the distant future, observers of the final moments (if they existed, which is hard to imagine) would be able to see the backs of their heads. They would vanish, crushed into nothingness, knowing that the cosmos was, after all, finite. With hope in their hearts (if they had hearts), they would die dreaming of a new cycle of existence, in which the energy they once were would find new ways of coalescing into complex material forms, some of them able to contemplate the meaning of infinity.

Another hope is that the shape of the cosmos will be calculable unequivocally from fundamental theory, when the union between quantum mechanics and general relativity finally takes hold. For one of the great challenges of modern physics is precisely how to transcend the difficulties imposed when singularities are reached, be they at the beginning of time, as with the Big Bang, or at the final stages of stellar collapse, as with black holes. In both cases, we describe the physics with Einstein's general theory of relativity, knowing only too well that the theory breaks down for very small distances and/or very high matter densities. What are we to do? The only reasonable way out is to reach toward the theory of physics that describes, and very successfully at that, the physics of the very small and attempt to apply it to situations with highly curved spaces and overdense matter. Indeed, quantum theory seems perfect for the task at hand, since it naturally provides a small-distance cutoff, a limit to how deeply we can probe smallness, a consequence of Heisenberg's Uncertainty Principle.

The idea, which we will explore in great detail in Part 2, is that an observer interested in measuring the position of an object with increasingly large precision will eventually hit a hard brick wall beyond which she cannot gain information. In other words, quantum theory implies that there is a natural fuzziness to matter, a finite "smallness" to all small things. Things may be small, but not smaller than that. There are no such things as "point" particles in Nature, since any kind of material structure must dissolve into quantum uncertainty and fill a volume. In a sense, this smallest volume is the ultimate barrier between what we can know about the nature of physical reality and what we can't. Even more dramatically, in the quantum realm the very question of trying to know more, of trying to go beyond the limits imposed by uncertainty, doesn't make sense. This sort of logic, that there is a limit to how much we can know of Nature, drove Einstein crazy.

If we apply this notion to space, it is natural to expect that the same will be true: that there is a smallest distance of space beyond which nothing can be smaller. According to this view, space is not really a continuum but fuzzy, so that motion cannot proceed smoothly from point to point. If this is the case, there can't be a true singularity, since space cannot be squeezed to zero volume. This is the view followed by proposers of a quantum theory of gravity, of which Abhay Ashtekar, Lee Smolin, Martin Bojowald, and many others are active defenders. There is an intrinsic assumption here, which is that we can extrapolate the uncertainty limits of quantum mechanics, which apply to material objects and light, to space and time, which are conceptual tools used to describe material objects and their motions. Is this extrapolation justified?

A competing view is to consider that it is not space that needs to be "quiltarized" but the notion of point particles that needs to go. The idea is simple: if the smallest things that exist have some spatial extension, they can't be squeezed into nothing. That is precisely what quantum mechanics says, that material objects are both particle and wave (and neither), inasmuch as they have a spatial extension attached to them. So superstring theories propose that the smallest objects in Nature are not the electrons, quarks, and other particles we see in accelerators such as the Large Hadron Collider but one-dimensional lines of energy that can twist and wiggle in a variety of ways. Being line-like and often forming closed wiggly loops implies that these objects (strings) have extension and thus cannot be squeezed into a singular zero volume. Ergo, if superstrings dominated the dynamics of the early Universe, they couldn't be squeezed into a singular nothingness.

Superstring theories have the additional claim of being "theories of everything," in the sense that they potentially can offer a unified description of all particles of matter—as different vibrational patterns of the fundamental strings—and of the four forces of Nature—also described in terms of their particle carriers as vibrational modes of the string. I have presented a detailed critique of the notion of final unification and theories of everything in *A Tear at the Edge of Creation* and refer interested readers to it. Here I note that notions of final theories are incompatible with the scientific method. Given that we can only accrue scientific knowledge from measurements of natural processes, it is by definition impossible to be certain that we know all the forces of Nature or the fundamental particles that exist; at any point in time, new technological tools may reveal the new and unexpected and thus force a revision of our current knowledge. This all-encompassing, godlike vision of Nature is a romantic fantasy. At best, superstring theories or their heirs will be able to provide a complete theory of what is known of particles and their interactions *at the time of their inception* but never as the final word.[1] As an illustration, recall the far-future cosmologists trapped in a static and dark-island universe discussed a few pages back. What would their final theory be like?

It would be no doubt compelling to them, even if profoundly mistaken when viewed from our current perspective. How can we then be so sure that we are that much better off, that we are not missing a big chunk of the cosmic picture? Science is efficient at discovering what exists, if within reach, but it cannot rule out with any final authority what doesn't exist. Which takes us to the grandest question of all: Is our Universe the only one? Are there others, coexisting in some kind of timeless multiverse? If there is such a thing as a multiverse, how could we ever know? To address such questions, we must first examine what sort of fuel could possibly promote such a wild proliferation of universes. Accordingly, we will take a short detour into the realm of normal and metastable states of matter and how they might have affected the cosmic infancy.

CHAPTER 13

ROLLING DOWNHILL

*(Wherein we explain the notion of false vacuum energy,
how it relates to the famous Higgs boson, and how
it may fuel an accelerated cosmic expansion)*

Einstein's general relativity describes gravity as the curvature of space resulting from the presence of matter and energy. We don't know *why* matter (or, equivalently, energy) bends space, but we can use equations to compute *how*. Einstein's beautiful theory is just another level of pragmatic description. Certainly it's a step beyond Newton's action at a distance, since a bent space is an immediate thing, a local and not a faraway influence, but the cause is still obscure. If asked why matter bends space, Einstein would surely have responded that he didn't know. The theory rests on the so-called Equivalence Principle: that a mass will respond equally to the pull of gravity and to a force accelerating it. As long as the acceleration is the same, an observer (who can't receive information from the outside) is unable to discern its cause. Or, as Einstein liked to put it, an observer falling won't feel his own weight.[1] So far the Equivalence Principle has held strong, despite many tests.

As our current best description of the workings of gravity, Einstein's theory does make remarkable predictions. Given the observationally supported assumption that matter, when averaged over large distances, is distributed in a homogeneous and isotropic fashion (same in all of space and in all directions, the Cosmological Principle), the theory can make quantitative statements about the geometry of the cosmos as a whole. In practice, cosmologists do this by modeling matter and radiation as a homogeneous gas endowed with an energy density (that is, its mass and/or energy per volume) and pressure (the force the gas exerts per unit area, as when you blow up a balloon by pushing air into it). In Einstein's theory both the density *and* the pressure of the gas contribute to the curvature of space and thus to the dynamics of the cosmos.[2] For normal kinds of matter and radiation, both the energy density and the pressure give positive contributions to the equations modeling the evolution of a universe. The result is a universe that expands in time but with an expansion that slowly decelerates. Depending on the amount of matter, the universe may recollapse on itself or keep on expanding, but loses its oomph as it does, coasting to zero velocity infinitely far into the future. (The exception is for a universe with an open geometry, which will keep on expanding.) But "normal" is not always the rule in physics.

In general relativity, where the pressure influences the curvature of spacetime, strange things can happen: certain types of matter can have bizarre gravitational effects.

First, a quick excursion into kitchen physics. Water exists in three states: solid (ice), liquid, and vapor (gas). To make it go from one state to another, you change the temperature. So to turn liquid water into a solid, you put it in the freezer, where the temperature is below water's freezing point of 32 degrees Fahrenheit (or 0 degrees Celsius). You can say that water in a liquid state inside a freezer is not in its most natural state, so much so that it will transform, losing energy to the environment and slowly condensing into ice. We can say that liquid water in a freezer is in a "metastable" state, a state in which the energy is not as low as it could be. The change from a metastable to a stable state is called a "phase transition."[3] Can other kinds of matter go through phase transitions? Definitely!

It happens all the time, as long as the right temperature (and/or pressure) conditions are satisfied.

The same notion applies to particle physics. The particles of matter can also go through different phases, in which their properties change. For example, we currently exist in the normal phase of matter in which electrons are about two thousand times less massive than protons. Think of it as the "icy" phase of matter. However, at higher and higher energies, particles start to morph, and their masses eventually change to zero. Imagine that we could hold a chunk of such massless matter in our hands at current energies. Just like liquid water in the freezer, this chunk of massless electrons and protons (or better: quarks, the constituents of protons) would be in a metastable state. It wouldn't last for long, transitioning into the more familiar kinds of matter we see around us. Even though current accelerators cannot make such metastable chunks of massless matter, there is good reason to believe that this will be possible in the future. Just as with the invention of the freezer, this sort of technological development takes time and ingenuity. (And money—lots of it.)

There is, however, one place where this kind of metastable matter was found in abundance: the early Universe. As we go backwards in time, the cosmos was hotter, and energies were higher. Before one trillionth of a second after the initial Bang, the Universe was hot and dense enough for matter to be in this metastable state.[4] And here comes the amazing bit: matter in a metastable state contributes a *negative* amount of pressure to the equations ruling the cosmic expansion. And according to general relativity, negative pressure makes the expansion of a universe accelerate instead of decelerate. There is pent-up energy stored in a metastable state that forces space outwards. (In a mundane analogy, think of a mass connected to a compressed spring. If the mass is released, the pent-up energy in the spring will push the mass away. Negative pressure does something similar to the geometry of space.) The conclusion is remarkable: the early Universe could have undergone periods of accelerated expansion whenever matter found itself in a metastable state. The effect is so general that it doesn't even need a metastable state; cosmic acceleration will happen whenever matter is not in its "normal" state, that is, not in the state of lowest possible energy. A useful analogy is a ball on an incline. It will roll down until it finds a stable point where it can rest. Any point along the hill is a "displaced state" for the ball, and it will have extra energy compared to being on the ground. We can call this extra energy the "displaced energy." Likewise, a universe filled with matter in a displaced state will evolve with accelerated expansion until it "rolls" down to rest in its lowest energy state.

The attentive reader will recall that we have discussed accelerated expansion before, in connection with the cosmological constant. As long as matter remains in a displaced state (anywhere "uphill"), it has the effect of a cosmological constant. The key difference is that for the cosmological constant the ensuing cosmic acceleration is constant (hence the name), while for matter the acceleration can increase or decrease depending on how far from the normal state it is. The measure of how far matter is from its normal state is often called "false vacuum energy," although we will simply call it "displaced energy," since it is a measure of the excess energy from the normal state.[5] The larger the displaced energy, the faster the cosmic acceleration.

To complete the picture, we need one more ingredient: the culprit that enacts the changes in the properties of the particles, changing them from massless at high energies (high displaced states) to massive at low (the normal state). According to our current knowledge of particle physics, summarized in the so-called Standard Model, the culprit is another particle, the famous Higgs boson. In July 2012, scientists at the Large Hadron Collider announced that the Higgs boson had been finally discovered.

One way to picture the net effect of the Higgs is to imagine it as a sort of medium where all particles must move. This sounds a lot like the old electromagnetic aether, but not quite. The traditional aether was unchangeable and inert, while the Higgs can change and interact with ordinary matter. Like ordinary particles of matter, the Higgs also changes its properties at different temperatures. In fact, current models of particle physics use the changes in the properties of the Higgs to change properties of matter particles. Going back to the picture of the Higgs as a medium (like air or honey), at high temperatures the Higgs is essentially transparent, and matter goes through it without much trouble. This is the massless phase. At lower temperatures, the Higgs "thickens," and particles of matter traverse it with more difficulty. This difficulty acts like a viscosity and is interpreted as particles having a larger mass. This is why it is often said that the Higgs "gives mass" to the particles.

We still need to explain why quarks, the electron, and the other particles of the Standard Model have different masses from each other. The reason is that they feel the presence of the Higgs with different intensities. The more a particle "feels" the Higgs, the higher its mass in the normal phase. In the mathematical formulation of the Standard Model, this feeling is translated into the intensity with which each particle interacts with the Higgs. For example, the top quark—the heaviest quark and also the heaviest elementary particle known—has a mass 339,216 times larger than an electron's. We say that the top quark interacts more strongly with the Higgs than with the electron. The exception is the photon, which doesn't interact with the Higgs and thus remains massless.

Armed with the picture of the Higgs as a medium, we can essentially forget about the different particles that interact with it and simply imagine the Higgs as the ball that can roll uphill or down. When high uphill, the Higgs is far from its normal phase and thus carries a lot of displaced energy. A universe filled with the Higgs in this displaced phase will expand exponentially fast. As the Higgs rolls down toward its minimum energy value, the acceleration will decrease until it stops at the point when the Higgs reaches the minimum.

The simple image of balls going up and downhill and the relative amounts of displaced energy is behind the mind-boggling notion of the multiverse. We are now ready to explore it.

CHAPTER 14

COUNTING UNIVERSES

(Wherein the concept of the multiverse is introduced and its
physical and metaphysical implications explored)

The reader no doubt has noticed that I have been diligently distinguishing between "Universe" and "universe." At first, this may seem to be a small detail. However, current thinking in cosmology seriously considers the possibility that there is more than a single universe. This being the case, the distinction is necessary. I use Universe, with a capital *U*, to represent our visible expanse of space and all that exists within it, known and unknown. In other words, "Universe" stands for all that exists within our cosmic horizon, our causally connected home. As we have seen, the measured flatness of the curvature within our horizon provides extremely strong evidence that space may very well continue beyond that— possibly to spatial infinity, even if anything infinite is clearly inaccessible to our instruments. It would be thus tempting to extend our definition and call Universe the possibly infinite expanse of space beyond the cosmic horizon. But to be strict and consistent with the motto "We only know what we measure," I will stick to it. As a consequence, a possibly infinite universe may contain our Universe. More amazingly, there may be other universes "out there," a huge number of them.

107

According to the *Oxford English Dictionary*, a "universe" is "all existing matter, space, time, energy, etc., regarded collectively, esp. as constituting a systematic or ordered whole; the whole of creation, the cosmos." The use of the opening qualifier "all existing" complicates things a bit. If by "all existing" we truly mean all, the OED would include in its definition of universe all other expanses of space that may exist but are separated from ours by the joint barriers of space and time. In this case, there would be a single universe, and any expanse of space, including ours, would be part of it. However, if we search the word "multiverse," we discover a somewhat confusing definition: "A hypothetical space or realm of being consisting of a number of universes, of which our universe is only one."[1] So, if a multiverse exists, a universe cannot be "all existing matter, space, time, energy, etc., regarded collectively." The multiverse, rather, would be this "all existing matter, space, time, energy, etc., regarded collectively," and a universe would be a part of it, one of the many, possibly infinitely many, coexisting "island universes." The complicating factor is that a universe, even if part of a multiverse, could be spatially infinite. So the infinite fits into a grander infinite, somewhat like $\aleph_0$ "fits" into $\aleph_1$. In modern cosmology, as in mathematics, there may be different kinds of infinites.

Before we go any further, let me explain how such a thing as a collection of different universes, some of them possibly infinite, even starts to make sense. To make things easier to visualize, let's stay in two spatial dimensions. Consider a flat tabletop. Using your imagination, make it stretch out very far away into its two directions (north-south, east-west). If the stretching goes on forever, the tabletop is a flat infinite space. Little, amoeba-like, flat critters may live in this infinite two-dimensional universe. Now consider two tabletops, parallel to each other but not touching. Make the second tabletop also infinite and have other kinds of critters live there. (This is all happening in your head.) Imagine further that a narrow tunnel connects the two spaces somewhere. Now we have two infinite spaces connected by a narrow tunnel. Critters in each space, without access to the tunnel, believe they live in a single infinite universe. This is especially true if the tunnel lies outside the critters' cosmic horizon. They will never know that their universes are part of a larger structure, a two-dimensional multiverse. You could easily imagine a huge number of two-dimensional flat spaces stacked on top of one another, each connected to the next by a similar tunnel, each tunnel inaccessible to any inhabitants. Keep on stacking up and down toward infinity, and you just built an infinite two-dimensional multiverse in your head!

The stacked multiverse need not be so simple. Universes can actually be curved and finite, sprouting from a "mother" universe, which is itself infinite. Sprouting universes may also be infinite. Think of bubbles being blown from a piece of bubble gum. As anyone who has blown bubble gum bubbles knows, little bubbles will shrink back, while bigger ones may keep on growing (until they pop, but we are neglecting bubble popping here). Imagine that a bubble starts growing in a region of flat space heavily populated with critters. Some of them will be carried into the bubble, while others will remain outside, horrified to see their friends sucked into "oblivion." Happily, most critters in the growing bubble survive their ordeal and start exploring their world. Generations pass. Their scientists measure the curvature of space and see that their universe is closed, like the surface of a sphere. By now, because the bubble has kept on growing, the tunnellike aperture to the original universe is well beyond their cosmic horizon. They live in a closed, expanding universe, unaware of their connection to a flat, infinite space. An ancient creation myth explains that a god blew their bubble universe from a large universe where other gods live and pass their infinite time blowing countless universes. Meanwhile, critters in the original space saw the aperture to the bubble universe close more and more until it narrowed so much that it became impossible to cross. All that is left is a scar in space, marking the long-forgotten birthing event. Even if still connected as if by an umbilical cord, the bubble universe became isolated from the mother universe.

Could something like this actually exist? The amazing answer is yes, at least in theory. Here is how.

To begin, consider a universe filled with Higgs-like matter. (No need to think that it is the same as the Standard Model Higgs boson. Theories of particle physics that attempt to explain physics at energies beyond the Standard Model naturally include extra Higgs-like matter.) Let's start calling the Higgs-like matter by its proper name: field. The concept of field is absolutely key in modern physics, having made a triumphal entry with Michael Faraday and James Clerk Maxwell's theory of electromagnetism in the mid-nineteenth century. Essentially, a field portrays the spatial influence of a certain source. For example, you can construct the temperature field in a room by simply measuring the temperature at different spots. This field, which only depends on the value of the temperature in a point in space, is called a *scalar* field. Another type of field is the velocity of water flowing in a river. Unless the flow is perfectly uniform, there will be eddies and deviations from the underlying flow. This kind of field, where not only the value at a point in space but also the direction at that point is important, is called a *vector* field. The wind flow around a house is another example of a vector field. Higgs-like fields are scalar fields, while the electromagnetic field is constructed from both a scalar and a vector field.

Back to our model universe. Consider further that this cosmic-filling, Higgs-like field is away from its minimum-energy state so that its displaced energy drives the universe into an accelerated expansion. Here enters the essential point that led to the multiverse idea: there is no need for the whole universe to be filled with the displaced energy of a scalar field; a small patch will do, and as long as it is large enough, it will inflate like a balloon in the background of the large, possibly infinite universe. This is similar to the image of the growing bubble in the stretched piece of bubble gum! Here the "push" to make a bubble grow doesn't come from a god but from the displaced energy of a scalar field. How big need the patch be in order to inflate exponentially fast? Not big at all, of the order of the cosmic horizon at the time it happens. For example, at the energy scale that the Higgs drives the particles of Nature into having a mass (at about one trillionth of a second after the Bang), the size of the patch would be around a millimeter or so. The closer to the origin of time you go, the smaller the patch needs to be. We are thus assuming that a given patch of space filled with a scalar field originated in the past, just the way our Universe did. We will get to the question of this origin soon.

We can thus imagine a scenario where a vast expanse of space is filled with a scalar field in a quilt-like pattern, so that different regions have the field at different values from its lowest energy state. (Imagine each region having its own ball on a hill, but with each ball at a different height.) Regions where the patch of space is large enough will expand exponentially fast, with a rate determined by the amount of displaced energy in that patch. (The higher the ball, the faster the expansion.) Space will quickly fragment into a plethora of growing bubbles, each expanding at its own rate, each a potential universe connected to the mother universe by a tube of space similar to an umbilical cord, usually called a "wormhole." This scenario, called "chaotic inflation," was proposed in the early 1980s by the Russian-American cosmologist Andrei Linde, now at Stanford University. "Chaotic" here refers to the random distribution of values of the scalar field at different patches of space.

Linde added a fascinating twist to his model: quantum mechanics teaches us that in Nature nothing sits still. Everything vibrates, even if these vibrations are imperceptible at the large scales of our everyday life. For the scalar field populating our hypothetical universe, however, this quantum jitteriness is very important; the farther from the minimum of energy the field is, the stronger its quantum jitteriness. If, in an already inflating bubble, a large enough chunk of the field is kicked uphill into a higher energy spot, it will start growing at a different rate. As a result, it will sprout away from its mother universe, becoming a universe of its own and a grandchild of the original. The reader can picture what might happen: bubbles will sprout from bubbles in endless fashion, each a potential universe with its own history. Linde's conclusion was that a universe filled with a scalar field displaced from its minimum energy will necessarily lead to a sprouting multitude of growing universes, a multiverse with no beginning or end.

Meanwhile, another Russian-American cosmologist, Alexander Vilenkin from Tufts University, proposed an alternative theory with similar consequences. Vilenkin looked at fields with an energy hill that started out very flat, as if on top of a mountainous plateau. The same way that quantum effects would kick Linde's field up and downhill, in Vilenkin's model the field would be kicked randomly in this or that direction along the plateau in different regions of space. As long as the patch was large enough, it would grow exponentially fast, creating a similar plethora of sprouting bubbles. Vilenkin called his model "eternal inflation," since he concluded that there would always be patches of the field high enough along the hill to create inflating regions: if in some regions the field would end up rolling down and the phase of accelerated expansion would end— as presumably happened in our own Universe—in others it will be just beginning. In fact, Vilenkin showed that inflating patches multiply faster than they decay.[2] So my two Russian-American colleagues, both delightful company in meetings, came up with bizarre models of an eternally reproducing multiverse: even though each growing patch has a history with a beginning, the multiverse itself may be eternal. Our Big Bang would be a local occurrence in a vast collection of possible cosmic histories.

Can such an idea, crazy as it sounds, be actually physics and not mere speculation? Any scientific hypothesis must be testable. Experiments or otherwise observational data must be collected to consider its scientific viability. Given that we have no evidence that we live in a multiverse—and direct evidence may well be impossible to gather—we must consider the idea with great care, checking what evidence we do have in hand, and add to it what we may collect in the future, if anything.

First of all, let's examine the notion of accelerated cosmic expansion. Is it reasonable? Absolutely! Since 1998, we have very convincing evidence that we do live in a dark energy–driven accelerating Universe. This is a most remarkable finding that should not be overlooked. It is even more remarkable when we consider that this accelerated expansion only started about five billion years ago. In other words, accelerated cosmic expansions are not only real but have a beginning and, possibly, an end. Five billion years is a peculiar time, coinciding with the formation of our own solar system. This is sometimes called the "coincidence problem": Why did the cosmic acceleration start then and not much earlier or later?

Another strong motivation for an epoch of accelerated expansion is the inflationary model of cosmology, proposed in 1981 by American cosmologist Alan Guth. This is the original model of cosmic inflation, which later influenced both Linde's chaotic inflation and Vilenkin's eternal inflation. Guth was concerned with some questions that the standard Big Bang model—which describes the Universe as emerging from a hot primordial soup of matter and radiation 13.8 billion years ago—couldn't answer. For one thing, why should the cosmic geometry be so flat? Why not closed or open? For another, the temperature of the cosmic microwave background, the same across the Universe to one part in one hundred thousand, seemed remarkably fine-tuned. The size of the cosmic horizon at decoupling would not have allowed for this confabulation: to have the same temperature now, particles and photons at decoupling needed to have interacted over distances much larger than their horizon allowed. That's how temperature is regularized in the Universe or in a bathtub, by particles (or water molecules) colliding with one another. The same way that the larger the bathtub, the longer it would take for water to reach the same temperature throughout, radiation in an expanding Universe takes time to regularize its temperature. And there simply wasn't enough time for this to happen since decoupling. This being the case, how could the photons in opposite directions in the sky know what temperature to be at?

Guth proposed that the young Universe went through a short-lived ultrafast period of accelerated expansion that he called "inflation." His idea was similar to the ball uphill discussed above: a Higgs-like field would be trapped in a metastable state, and as long as it remained there, the Universe would expand exponentially fast. Both Andrei Linde and Andreas Albrecht of the University of California at Davis, with Paul Steinhardt of Princeton University, soon realized that Guth's model had an issue called the "graceful exit" problem: the field would remain trapped too long to emulate the Universe we see. They independently suggested that if the energy profile for the field had a flat plateau all would be well. This is the model that inspired Vilenkin's eternal inflation scenario.

Inflation naturally explains why the Universe is so flat: a small patch that got stretched by dozens of orders of magnitude would appear flat, since now it would be part of a huge spherical surface. In this case, our cosmic horizon would be a small section of a much larger universe inaccessible to observations. Inflation also explains why the microwave background has the same temperature to such high accuracy. Essentially, since our whole cosmic horizon originated from the same blown-up initial patch, it is natural to expect that the particles and photons would share their thermal properties.

Inflation goes one step further. Recall the small quantum kicks that feed the endlessly sprouting cosmoids? Well, the same kicks that cause the field to go up and downhill within each patch give rise to small fluctuations of energy; like the uneven surface of a lake, some spots within the expanding patch have a bit more energy, others a bit less. With inflation, those tiny quantum inhomogeneities get blown up to astronomically large lumps of field stuff. Fast-forward to around the time of decoupling, when hydrogen starts being produced. Since gravity is an attractive force, these overdense and underdense regions will attract more or less matter, making it concentrate in some spots like rainwater collecting in puddles here and there. Ultimately, the overdense regions are responsible for creating the observed distribution of galaxies and clusters of galaxies in the Universe. Space would be like a bad dirt road, with dips and bumps: less matter collects near the bumps, and more falls into the dips. In other words, inflation offers a mechanism to explain how galaxies and clusters of galaxies were born. Since photons also feel the dips and bumps in space, inflation predicts that these inhomogeneities would have left an imprint in the cosmic microwave background in the form of tiny temperature fluctuations (hot and cold spots). This checkered temperature map has been measured to great accuracy by the Wilkinson Microwave Anisotropy Probe satellite and with even finer accuracy by the European Planck satellite. Amazingly, the data agrees to high precision with what some models of inflation predict, giving cosmologists confidence that the young Universe did go through something like an accelerated period of expansion as modeled by the inflationary theory.

If this is the case, and if our cosmic horizon is very nearly flat, it follows that the universe must be much bigger than what we can see, extending well beyond the measurable Universe. Even though no definite statements can be made about the existence of infinite quantities in Nature, the universe is probably extremely large, possibly even infinite. It is then reasonable to speculate that other inflating regions may exist, as predicted in eternal inflation models.

A key ingredient of inflation is, of course, the scalar field. Can we be sure that such a field existed during the early stages of cosmic evolution? No—at least not yet. However, the success of the Standard Model of particle physics and the recent discovery of the Higgs boson make credible the hypothesis that at higher energies something like the Higgs, or something that has the same net effect on the cosmic expansion, could have been present. A huge number of models attempting to expand current knowledge of particle physics beyond the Standard Model invoke scalar fields. Superstring theory, for one, comes out with many possibilities. Even if you are not an enthusiast of theories invoking supersymmetry, we can still be fairly confident that new physics will take hold at energies higher than what we can probe at present, potentially offering viable candidates for the inflation-driving field or fields.

Healthy science needs a combination of humility and hope: humility to accept the extent of our ignorance and hope that new discoveries will illuminate the current darkness. However, when we are at the edge of knowledge and data is not forthcoming, well-grounded speculation is the only strategy at our disposal. Without imagination science stagnates.

I would be remiss if I didn't include here a brief discussion of how superstring theory contributes to the notion of a multiverse. Several colleagues, all devoted to superstring theory, have written popular accounts that I list in the bibliography. Of these, I highlight those by Brian Greene and Leonard Susskind, in case the reader wants to learn more. For our purposes, however, what I will say next should suffice.[3]

CHAPTER 15

INTERLUDE: A PROMENADE ALONG
THE STRING LANDSCAPE

(Wherein the notion of the string landscape is discussed,
together with its anthropic motivation)

I n order to make mathematical sense, superstrings must live in
spaces with more than the usual three dimensions. This creates
an immediate challenge to the theory, as it must somehow explain
why, if that's the case, we only see three of the dimensions. We
also need to know how many extra dimensions exist. Is it one, two,
twenty? Here the theory invokes a hypothetical new symmetry of
Nature: supersymmetry. We met it briefly when discussing dark
matter. In this chapter I'll go into a little more detail. The way we
see it, there are two kinds of particles in Nature, those composing
matter (electrons, quarks, and a few others) and those transmitting
forces (photons for electromagnetism, "gravitons" for gravity, less
familiar ones called gluons that keep the quarks locked within protons and neutrons, and the heavy Z^0 and W^+ and W^- of the weak
nuclear force, responsible for radioactive decay). Supersymmetry
says that particles of matter may be converted into particles of
force and vice versa. The net result is that each particle has a supersymmetric partner: the electron, for example, has a "selectron"; the
six quarks have "squarks" and so on.

117

You may be wondering why on Earth doubling up the number of elementary particles in Nature is a good idea. The answer, the original motivation for supersymmetry in the mid-1970s, is that theories with supersymmetry could explain why the energy of empty space is zero. If it weren't, if there were some kind of residual energy in space, it would act just like a cosmological constant, accelerating the Universe. In the mid-1970s that was a no-no, since there was no evidence for a cosmic acceleration. (The discovery of dark energy only happened in 1998. Before then, everyone expected the cosmological constant to be zero, and supersymmetry offered a way to explain how it could be so.)

The problem was the *vacuum*, what physicists call the state without any particles. The vacuum is as close to "nothing" as physics gets: it is empty space. However, quantum physics complicates things. As the reader may recall, the essential property of quantum physics is that everything fluctuates: a particle's position, its velocity, its energy. So even if empty space has zero energy, quantum fluctuations around zero energy may push the energy up here and there. In regions where there is an excess of energy, particles can pop up for a while as the excess energy is converted into matter via $E = mc^2$, before they disappear back into "nothing," like bubbles in a boiling soup. When the tiny quantum effect is summed over all of space, it generates a *huge* net contribution to the energy. Supersymmetry suppresses such fluctuations, making the energy of empty space vanish in some models. It would thus help explain why the cosmological constant is zero. But now, with dark energy, we know it isn't! Since 1998, the motivation to use supersymmetry to cancel vacuum fluctuations is not so strong anymore.

Explaining small numbers is very difficult in physics. Still, there are other advantages to supersymmetry, such as explaining why the scale in which particles get their masses from the Higgs is so much smaller than the scale in which spacetime starts to fluctuate due to quantum effects (by some sixteen orders of magnitude) or to provide viable candidates for dark matter. For these reasons, although there is no evidence that supersymmetry exists (experiments hunting for particles predicted by many supersymmetric models have so far found nothing), it still remains strong as a possibility in the minds of many physicists.

Back to strings. Once you combine them with supersymmetry, the number of extra spatial dimensions is fixed: superstrings can only live in nine spatial dimensions. Five possible string theories have been found, and mathematical physicist Edward Witten from the Institute for Advanced Study has shown that they are all related in a theory with yet another extra dimension called "M-theory."[1]

So if strings are a theory of Nature, six extra dimensions must be invisible to us. How can this be the case? My doctorate thesis and some of my work during my years as a postdoctoral fellow were dedicated to this question, which then was very new. In particular, if we attempt to combine superstring theory with Big Bang cosmology, how to explain that three dimensions grew while the other six remained small? Numerous models were proposed, most using some kind of attractive interaction between the particles generated by string vibrations that would keep the extra dimensions very small. Since we don't have a viable candidate for a superstring theory, we still don't know how to answer this question. Indeed, new ideas relax the smallness of the extra dimensions in favor of large (but still unobservable) ones. Others have proposed that the extra dimensions are very large, and we live in a space (a brane) embedded in them, like a slice of toast hanging in open air. Lisa Randall, a friend from those early years of extradimensional cosmology and the first woman tenured in theoretical physics at Harvard University (sure took them a while), has written a book where she explains the brane concept, which she developed in collaboration with Raman Sundrum in 1999.[2]

Our interest in this discussion is that superstring theory also predicts the existence of a multiverse. It goes by the name of the "string landscape," and it encompasses all possible warpings and foldings that the extra six dimensions could in principle go through.

(Imagine deforming a ball of play dough into different shapes, with different numbers of holes, for example.) Each shape of the extradimensional space implies different physical properties in our three-dimensional reality. The string landscape is a space of the possible geometries that the extra six dimensions can have, not a space you can take a walk on. The invention of the landscape prompted an interesting change in the psychological attitude toward string theory. Originally, the main attraction of string theory was to provide *the* theory of Nature: its uniqueness was its main strength, its beauty. The hope was that once the fundamental equations of the theory could be solved, they would provide a *single* and inevitable solution: our Universe! Alas, things didn't go this way. Further study indicated that a huge number of solutions is possible, each a "dip" in the landscape, with estimates as large as 10^{500}, resulting from the topologically rich structure of the extra dimensional space. How could we pick one solution out of 10^{500}? What could have guided Nature to find a preferred choice, or "true vacuum"? So far, no one has come up with a compelling selection principle. In the absence of one, the long sought-for uniqueness of string theory is gone.

Furthermore, since each possible conformation of the extradimensional space translates into a different kind of four-dimensional spacetime geometry imbued with a unique set of particles and their interactions, different dips in the landscape (string "vacua") predict entirely different universes. Why would ours, with the measured values of the fundamental constants and rate of expansion, be special? Is it somewhat related to having a Universe with living beings in it? Since the seventeenth century, the trend of physical cosmology has been to show how unimportant we are in the big scheme of things. Would superstrings be a reversal of the Copernican revolution?

In 2000, Raphael Bousso of the University of California at Berkeley and Joseph Polchinski of the Kavli Institute for Theoretical Physics at University of California at Santa Barbara (where I once was a postdoctoral fellow) had the idea to combine the string landscape with eternal inflation. They reasoned that different dips (more technically, minima) of the string landscape would be separated by rapidly inflating regions so as to quickly become isolated from one another. In other words, *all* string vacua are in principle realized as separate universes, not only ours. This way Bousso and Polchinski brushed aside the question of our uniqueness. Bringing their idea even closer to models of inflation, they proposed that small quantum kicks can induce small changes in the geometry of the extra six-dimensional space, which in turn induce random motions along the landscape. The string landscape multiverse would thus consist of the mutating realizations of the multiple string vacua, each a different universe, each with its own particles (or none) and possibly even unique physical laws, although it's not clear at all to me how different physical laws appear in the superstring scenario, in spite of many such claims. (To change masses and strengths of interactions between particles does *not* lead to changing the fundamental laws of Nature, such as conservation of energy or of electric charge.)

The eternally inflating string multiverse hypothesizes that countless many universes exist "out there," unaware of one another. For the first time in the history of science, the unknowable gained the imprimatur of theoretical physics. To say the least, such radical departure from the time-honored methods of empirical science has raised many eyebrows and at least as many questions. In this mad plurality of randomly sprouting cosmoids, how can we ever hope to explain our own existence? Or should science lay this question to rest? Here a significant number of string theorists make use of a principle that, a few years back, would have been anathema to the whole philosophy of uniqueness that originally motivated the development of string theory: according to the Anthropic Principle, a subject of intense debate, our uniqueness is not predicted but postdicted.[3]

In the 1970s astrophysicist Brandon Carter suggested that we shouldn't be surprised to find ourselves in a universe conducive to life. After all, only a universe with the right properties (read: values of the masses and interaction strengths of the fundamental particles, plus a few cosmological parameters that ensure the cosmos is old enough and expands at the right clip) would allow for several generations of stars to exist and hence for life to have a chance of taking hold. In other words, the constants of Nature, such as the strength of the gravitational force or the mass of the electron, are the ones that allow for life. Given the fragility of the physical processes leading to the formation and burning of stars in an expanding universe, there isn't much leeway in the allowed values of these constants. We could only exist in the few rare universes where the constants of Nature have values very close to what we measure.

Bousso and Polchinski conjectured, and others followed, that the Anthropic Principle is the only way to have some kind of selection criterion to pull our Universe out of the amazingly vast string landscape. When one of the architects of string theory, Leonard Susskind of Stanford University, joined them in 2003, the landscape idea took off like a rocket. Anthropic reasoning states that we could only exist in vacua that allow for a small cosmological constant, like the one that seems to be behind dark energy. Since all vacua in the landscape are in principle realized somewhere in the multiverse, we shouldn't be surprised that ours is too, even if we occupy a rare corner of space. There is no uniqueness, just a profusion of possibilities, including our atypical one. Our Copernican mediocrity is thus fully restored. (Of course, if someone found a compelling reason for our vacuum to be the preferred one, the global minimum of the string landscape, overnight the whole anthropic fad would be quickly forgotten as a wrongheaded way to think about physics.)

Objectors to the usefulness of the Anthropic Principle, including myself, state that it doesn't really help us learn anything new, offering at most a range of plausible values for a given variable by retrofitting what we already know. Anthropic reasoning narrows possible choices of physical parameters based on the properties of the known Universe, but it doesn't offer a pathway to explain why this choice and not others. It accommodates without illuminating. Here is an illustration. Consider, for example, the median height of an adult American male, 1.77 meters (about 5 feet 9.5 inches). Simple statistics will tell you that during a walk along city streets the probability of finding a man with a height between 1.63 meters (5 feet 4 inches) and 1.90 meters (6 feet 3 inches) is 95 percent. This is what anthropic reasoning would give you, a range of heights based on knowing the median height. But if we didn't know the median height, anthropic reasoning wouldn't be able to tell us anything very useful. In particular, it wouldn't explain the most crucial number in this example, the median height of the adult American male, a much more complicated question that calls for a multidisciplinary study.[4]

Could something like a multiverse with various possible values of the constants of Nature appear within the context of eternal inflation without invoking superstrings? In principle, yes. We could imagine a theory with many scalar fields, each related to a different set of constants in the low-energy minimum, just as the Higgs determines the value of the masses in the low-energy vacuum in which we exist. In each realization of inflation, a horizon-size patch of space would have a family of scalar fields following different histories, each ending at different energy minima and thus generating different sets of constants. Or you could have a single scalar field or a few scalar fields with many possible energy minima; in different patches the field would roll to different minima, giving rise to different physical constants.

Taken together, the arguments above suggest that the multiverse is at least theoretically possible. For the sake of argument let's entertain the idea that we do live in a multiverse. Could we ever know? Can the multiverse be observed? In other words, is the multiverse a testable scientific hypothesis, or is it just idle theorizing, leading to a dangerous schism in the physics community? More to the point, is the multiverse knowable or unknowable?

$$\asymp\!\!\!\asymp$$

CAN WE TEST THE
MULTIVERSE HYPOTHESIS?

(Wherein we explore whether the multiverse is a proper
physical theory or mere speculation)

When it comes to far-out ideas, physicists must be ruthless. Throughout history, many wild speculations have been duly proposed and duly believed by many, only to be pummeled into oblivion by the sheer force of very convincing evidence: the electromagnetic aether, the phlogiston, the caloric, Le Verrier's planet Vulcan. Blame it on the excesses of human imagination combined with an inflated zest for an idea. After all, if you are not passionate about your idea, who would be? We want to know, we *need* to know, and we will do what we can to come up with an apparently rational description of an unexplained phenomenon. We will advance all sorts of compelling reasons as to why such a hypothesis must be correct. Of course, we learn by trying, and failed attempts at explanations bring us closer to explanations that do work. If you don't like failure, science is not for you. The Island of Knowledge doesn't grow in a monotonic, predictable fashion. It sometimes retreats back, exposing a gulf in our understanding that we thought we had bridged. Imagination is essential to this process of invention and discovery, but it can't work alone: testability is the cornerstone of any scientific theory building. Twenty theoretical physicists locked up in an isolated room would come up with a universe very different from the one we live in.

The notion of a multiverse poses a serious threat to this modus operandi. If other universes exist outside our cosmic horizon, we can't *ever* receive any signal from them or send anything to them. If they exist, they occupy a realm entirely inaccessible to us or to our instruments. We could never see or visit them, or be seen or visited by observers who may live in them. So, in a strict sense, the existence of the multiverse can never be directly confirmed. Cosmologist George Ellis, from the University of Cape Town in South Africa, forcefully defends this position: "All the parallel universes lie outside our horizon and remain beyond our capacity to see, now or ever, no matter how technology evolves. In fact, they are too far away to have had any influence on our universe whatsoever. That is why none of the claims made by multiverse enthusiasts can be directly substantiated."[1]

Few physicists nowadays would follow the old positivist banner, expressed so dramatically when the Austrian philosopher and physicist Ernst Mach stated as late as 1900 that atoms couldn't exist because they couldn't be seen. (Sadly, Mach stuck to his anti-Atomism until his death in 1916.) There are different ways we can infer that something exists, even if we can't see or touch it. Astrophysicists do this when they deduce the existence of a massive black hole in the center of our galaxy from the motions of stars nearby, and extend this conclusion to other galaxies. Particle physicists do this when they obtain a particle's properties from the tracks it leaves in a detector. No one "sees" an electron—only the tracks electrons leave in various devices. We conclude that electrons exist from the clues they imprint in our machines. "Exist" may be too strong a verb; we construct the idea of an electron to make sense of the blips and lines we collect with our tools when we probe the world of elementary particles. Likewise, we construct the idea of dark energy as an economical explanation for the redshifted spectroscopic signatures of distant objects.

So the question is not whether we can hope to "see" a neighboring universe, but whether there is any way to detect its presence indirectly within our cosmic horizon. This would *not* be a test for the existence of the multiverse, since it would only indicate the possibility that neighboring universes exist. But it would surely offer some support to the idea, and it is a very attractive line of research. The distinction between finding observational signatures of neighboring universes and finding those of the full-blown multiverse is *very* important, as the two are often confused. So I stress it again: even if convincing observational signatures of neighboring universes could be found within our cosmic horizon, they would *not* confirm the existence of the multiverse. Although for some physicists detecting the existence of another universe makes it natural to generalize to a multiverse, the data would not support this conceptual jump. Two or three neighboring homes don't make a country. The existence of a multiverse, even if not of infinite extent, is an unknowable.[2]

As our previous discussion of Big Bang cosmology has indicated, the best current observational tool at our disposal to explore properties of the Universe is the cosmic microwave background. Could other universes have somehow left an imprint in the background of photons drifting across space for the past 13.8 billion years?

"When Universes Collide" would be the appropriate title for an article on the topic.[3] Could a neighboring universe have collided with ours in the past? Clearly, if this happened, the collision wasn't very violent; otherwise we wouldn't be here speculating about it. But yes, nearby universes could collide with ours as they grow and stretch away from one another in the multiverse. (Touching is more reasonable than colliding, which sounds destructive and violent.) In 2007, Alan Guth, together with Alex Vilenkin and Jaume Garriga, from the University of Barcelona, speculated that this would indeed be the case. Just as the naïve picture of two colliding soap bubbles would suggest, the collisions would create wiggling vibrations on their surfaces. These, in turn, would reverberate inwards in a disklike fashion, making whatever exists within the bubble jiggle along. The collision would create ripples in the space geometry within each of the bubbles, and those would propagate across space, like waves in a pool that make people and objects go up and down. The interesting aspect of these ripples is that they would be disklike, creating concentric rings. The microwave sky maps would then show ringlike patterns imprinted when the collisions took place.

Several cosmologists, including Anthony Aguirre from the University of California at Santa Cruz and Matthew Kleban from New York University and their collaborators, have constructed theoretical scenarios of what would be seen if such collisions did happen in the past. Ringlike temperature fluctuations could appear in the photons of the cosmic microwave background with a variety of sizes and intensities, depending on the details of the collision. In addition, the photons could also display a polarization pattern, that is, be aligned in a particular direction in the sky, like dominoes standing on end.[4] A preliminary search based on data from the WMAP satellite has turned up a negative result. But the issue is not yet decided. The Planck satellite team is expected to release data that may carry the signature Kleban and others have been waiting for: a disk-shaped circular pattern in the microwave background, with photons displaying a double-peak polarization signature in a particular direction at the edge of the disk. This signature would be quite unique and provide strong evidence that such a collision did indeed happen in the distant past; it would be hard to justify it in any other way.

Note, however, that we wouldn't be able to learn much about the physics operating in this neighboring universe, that is, the kinds of matter and forces that exist there, or whether the laws of Nature there are the same as ours, although the calculation for the collision rate does assume similar laws of Nature there, at least generally speaking. As if brushed by a ghost, we would know of an alternative reality beyond our Universe, tantalizingly close yet unreachable, existing yet unknowable. Even if the superstring landscape scenario would get some kind of indirect confirmation from particle physics, and hence lend further credence to the multiverse hypothesis, we wouldn't know which of the huge number of possible cosmoids brushed against ours in the past, or whether another such collision could happen in the future, possibly leading to our demise. Like the legendary seeker who after facing great perils discovers a dark secret of great destructive power, detection of a neighboring universe would lead to a mix of triumphal accomplishment and primal fear. It is only fitting that the patterns in the sky are ringlike, evoking fabled rings capable of much power and destruction, such as Richard Wagner's Ring of the Nibelung or J. R. R. Tolkien's Ring of Lord Sauron.

Although the chance of detecting such a pattern in the cosmic microwave background is exceedingly small, Aguirre, Kleban, and others are making the important point that the existence of other universes, despite being a seemingly esoteric possibility, is within the purview of testable physics. As is often the case with exotic research topics, even if the chances of success are small, the payoff is so enormous that the effort is well worth it. However, I stress again that even a positive detection of a neighboring universe would not prove the existence of a multiverse. Within the present formulation of physics the multiverse hypothesis is untestable, however compelling it may be. Extrapolation from two universes or even a few to many—possibly infinitely many—is not automatic.

More to the point, the notion of "infinitely many" is not testable as a matter of principle: to know of infinite space, we would need to receive signals from infinitely far away; to know of infinite time, we would need to receive signals from the infinitely distant past; to know if the universe will expand forever, we would need to monitor its expansion forever, given that we can't predict whether some new physics would halt or reverse the expansion at some point in the future. Even though the notion of the infinite has enormous mathematical appeal and naturalness, we can never be certain if it is ever realized in Nature. In the physical world the infinite is an unknowable. The most that we can do is speculate about its existence from within the finite shores of our Island of Knowledge.

The inflationary hypothesis and the possible existence of the multiverse stretch the notion of testability in physics to the breaking point. We have seen why this is the case for the multiverse, which, strictly speaking, is untestable. The case with inflation is subtler. Inflationary cosmology, taken at its most model-independent formulation possible, does make several predictions that have been verified: chief among them are the flatness of the Universe and the temperature homogeneity and isotropy of the cosmic microwave background. But we must remember that these are not truly predictions: inflation was *designed* to solve the flatness and the horizon problems from standard Big Bang cosmology. It is thus no big surprise that it does so.

Where inflation does add new predictions is in the shape of the temperature fluctuations about the homogeneous background of cosmic microwave photons. Like tiny waves on the surface of a lake, inflationary cosmology states that these fluctuations had their origin in the quantum jitteriness of the scalar field responsible for inflation. During inflation these tiny lumps are greatly stretched, eventually even beyond the cosmic horizon. As the Universe keeps on expanding, some of these fluctuations "come back" inside the cosmic horizon, now with sizes that reach astronomical scales. Wherever there is an excess of energy, gravity will cause matter—meaning mostly hydrogen atoms—to collect more. Likewise, the photons from the microwave background will slide toward these lumpier regions, gaining energy (i.e., increasing their temperature) as they do so. This motion will translate into tiny temperature fluctuations in the microwave background. After millions of years, the matter collected into these lumpier regions will coalesce into the first stars and then galaxies. A great triumph of inflationary cosmology is thus to offer a causal mechanism to explain where galaxies come from and why they are distributed in space the way they are, in a hierarchy of clusters that resembles the froth in a bubble bath.[5]

The temperature variations in the microwave photons measured with current satellite technology and ground-based detectors constitute a registry of the primordial matter fluctuations. To study them is to open a window into the earliest moments of time. In spectacular fashion, inflation connects the quantum with the astronomical. As measurements become more precise, specific models of inflation can be—and some have been—ruled out. An additional signature of inflation is the spectrum of the fluctuations in the spacetime geometry: if the matter concentration fluctuates in certain ways, so will the geometry of space about it. Inflation will also stretch these spatial fluctuations and produce what is called a flat spectrum of "gravitational waves." These also leave an imprint on the microwave background, somewhat similar in nature (but not in kind) to the polarization fluctuations from hypothetical collisions with neighboring universes. The hope is that the Planck satellite will also be measuring (or not) this polarization spectrum. If it does, and if it has the expected signature, we can be fairly certain that something with the broad outlines of inflation did happen when the cosmos was young.*

Still, it is one thing to confirm the general outlines of a phenomenon and another to test its detailed formulation. Inflation leaves many questions unanswered. Although data is helping us narrow down the possibilities, we won't be able to discern from current or near-future observations exactly what caused inflation. Was it a scalar field? If so, what sort of ultra-high-energy physics originated it? Inflation also leaves unanswered the details of the all-important transition from a rapidly expanding patch of space to the slower

---

*As this manuscript was going to press, results from the experiment BICEP2 announced on March 17, 2014, offered a first glimpse at the reality of this scenario. Inflation, and the stretching of quantum fluctuations, has received convincing empirical validation.

expansion pace that dominated cosmic history for the first five billion years or so. Supposedly, it was during this transition at the end of inflation that the Universe got hot, as the energy stored in the scalar field rolling downhill morphed explosively into other types of matter, possibly even the electrons and quarks that we are familiar with. In fact, many cosmologists now refer to this explosive particle creation as the true Big Bang. But in spite of many attempts, including some of my own, we have nothing more than a preliminary understanding of how this transition happened or what kinds of particles were there. The main obstacle is our total ignorance of the kinds of matter that existed at that early time, which correspond to energies trillions of times above what we can now test in the laboratory. Astronomical observations can rule out some of the models theorists come up with while constraining others, but they cannot isolate the one that actually corresponds to what happened. We can only know what is wrong, not what is right, a situation that would certainly get the nod from philosopher Karl Popper, who proposed that physical theories cannot be demonstrated to be right in a final sense, only to be wrong.

The best that we can do with inflation is to construct a workable model, consistent with all the measured parameters. As with Ptolemy's epicycles, however, the model may be just that: a fantastic concoction that "works." While it may fool many into believing it to be the real thing, its true worth is in providing an efficient summary of our current knowledge of early cosmic history.

Our next task is to address the greatest question of them all, the origin of the Universe. For neither cosmic inflation nor the multiverse brings us any closer to an understanding of the ultimate origin of all things. Indeed, in order to fully address the question, we must investigate the properties of matter and the quantum laws that dictate them. After all, consistent with the notion that the Universe has been expanding since its beginnings is the notion that in the distant past it was very small—so small, in fact, that the rules of quantum physics had to determine its overall behavior. As we will discover, however, such rules force us to abandon some of our most cherished notions of what we lightheartedly call "reality," in favor of a much more nuanced and mysterious description of the world and thus of the quantum Universe and our relation with it.

In quantum physics we will encounter head-on the two fundamental kinds of limits to knowledge that we explored so far: those imposed by the limited precision of our exploratory tools and those that are an essential consequence of how Nature works and that impose insurmountable barriers to how much we can know about the world and about the nature of reality.

PART II

FROM ALCHEMY TO THE QUANTUM: THE ELUSIVE NATURE OF REALITY

In reality we know nothing; for truth is in the depths.
—DEMOCRITUS, FRAGMENT 40

*And among such various and strange Transmutations,
why not Nature change Bodies into Light, and Light into
Bodies?*
—ISAAC NEWTON, *OPTICKS* (1704)

*If one wants to consider the quantum theory as final (in
principle), then one must believe that a more complete
description would be useless because there would be no
laws for it. If that were so then physics could only claim
the interest of shopkeepers and engineers; the whole
thing would be a wretched bungle.*
—ALBERT EINSTEIN TO ERWIN SCHRÖDINGER,
LETTER OF DECEMBER 22, 1950

*Each piece, or part, of the whole nature is always an
approximation to the complete truth, or the complete
truth so far as we know it. In fact, everything we know
is only some kind of approximation, because we know
that we do not know all the laws as yet. Therefore,
things must be learned only to be unlearned again or,
more likely, to be corrected.*
—RICHARD FEYNMAN,
FEYNMAN LECTURES ON PHYSICS

CHAPTER 17

EVERYTHING FLOATS IN
NOTHINGNESS

(Wherein we explore the Greek notion of Atomism)

What is the stuff that makes up the myriad things of the world, with their shapes, textures, and colors? Why is it that your skin feels so different from the pages of a book, a handful of sand, a burning fire, or a gust of cold wind? Why do substances change at different temperatures, and with such variation from substance to substance? To what extent can matter be morphed and engineered to serve our purposes? Is there such a thing as complete emptiness?

These questions are not modern. In Part 1 we encountered some of the Presocratic philosophers, the first to ponder the material nature of the cosmos. We have seen how Thales and his Ionian successors proposed, right at the inception of Western philosophy around 600 BCE, a unified theory of Nature whereby all stuffs of the world were but different manifestations of a primal substance, the embodiment of a reality always in flux.[1] To the Ionians, time was the essence of reality. In contrast, Parmenides and his followers proposed that the essence of Nature was not to be found in the transient but in the permanent; that what *is* cannot change, for it then becomes what is not. The truth, they argued, could not be ephemeral. To them, timelessness was the essence of reality. In the relatively short time span of one hundred years, philosophies of being and of becoming were proposed as mutually exclusive pathways to Nature's secrets.

A brilliant solution to this dichotomy came some two centuries after Thales, in the work of Leucippus and his prolific disciple Democritus. Instead of viewing being and becoming as two irreconcilable ways of thinking about reality, they proposed to see them as two ends of the same stick. For this, the duo conjectured that all things are made of small, indivisible bits of matter, the famous atoms, which in Greek means "that which cannot be cut."[2] The atoms were immutable, "what-is," and thus incarnations of being. They moved in the void, "what-is-not," a medium perfectly empty of matter. To the Atomists, both atoms and the void were equally fundamental to describe Nature. Parmenides would discredit the void as nonexistent, since in his philosophy that which-is-not cannot be: as soon as you say "there is a void," you are stating the void's existence, and if the void exists, it cannot not be.

The Atomists probably just shrugged at this excessive idealism and insisted that atoms moved in the void. Through their mutual combinations and mechanical rearrangements atoms would assume different shapes and forms, explaining the material diversity we see in Nature. Change, then, was simply due to the reordering of immutable atoms, like Lego pieces: being and becoming were joined together, in what could be considered another unified theory of Nature. At the same time, things had an identity, an essence that remained unchanged, despite possible transformations and decay. So the water that flows in rivers becomes clouds in the sky and water again when it rains; the seeds that grow into oaks reappear in mature trees and, in the course of time, will become trees again; worlds decay, and, from their remains, new worlds emerge. The dance of atoms animated Nature's permanent state of flux. Emboldened, Democritus went on to propose that the impact of atoms on our sensorial organs explained sensations as well: "by convention color, by convention sweet, by convention bitter, but in reality atoms and void."[3] Through his vast body of work, Democritus created a formidable explanatory device based exclusively on a materialistic description of reality. Still, he was wise enough to caution against the illusion of final knowledge: "In reality we know nothing; for truth is in the depths."[4]

Around 300 BCE Epicurus took up Atomism with renewed force and flair, refining some of the murkier ideas of his predecessors. He stated that atoms must always be invisible and that there should be an enormous variety of shapes made out of their combinations (akin to our modern molecules): "In addition, the indivisible, solid particles of matter, from which composite bodies are formed and into which such bodies are dissolved, exist in so many different shapes that the mind cannot grasp their number."[5]

More remarkably, Epicurus expanded the notion of multiple universes, *kosmoi*, that Leucippus and Democritus espoused before, suggesting that they are separated in space by well-defined boundaries: "the number of worlds, some like ours and some unlike, is also infinite." One could argue that these worlds (*kosmoi*) are simply other planets, but Epicurus makes it clear that he means a self-contained universe, or at least what we today would call a galaxy, separated from the rest by a spatial boundary: "A world (*kosmos*) is a circumscribed portion of the universe, which contains stars and earth and all other visible things, cut off from the infinite, and terminating in an exterior which may revolve or be at rest, and be round or triangular or any other shape whatever."[6] The notion of island universes or possibly even a multiverse is much older than commonly supposed.

Although the atoms of the ancient Greeks are different from those of their modern heirs, the notion that matter is made of indivisible blocks has been the driving force behind the physics of the very small ever since it was first proposed. In spite of its modern triumph, Atomism has a checkered history and in the West at least sank almost into oblivion during the Middle Ages. The situation only began to change during the early Renaissance, when some of the ancient Atomistic texts, most notably Lucretius's *On the Nature of Things*, were rescued from the dusty bookshelves of far-off European monasteries and private collections. As Steven Greenblatt argues in his excellent *The Swerve*, we owe the resurgence of Atomism and, more generally, of materialism to the intrepid early-fifteenth-century manuscript hunter Poggio Bracciolini, who found a copy of Lucretius's work among piles of half-forgotten scrolls in a German monastery.

There were, however, others who kept the Atomistic ideas quietly alive, if not so much in the old tradition of argumentative Greek philosophers, certainly as practitioners, as they attempted to tease matter into giving away its secrets through endless distillation and mixing. No account of our attempts to unveil Nature's mysteries would be complete without discussing the role of alchemy and its enormous influence on some of the patriarchs of modern science, such as Robert Boyle and Isaac Newton. To a large extent, alchemy represents the bridge between the old and the new, a concrete implementation of philosophical and spiritual beliefs into the scientific practice of experimentation. Its core notions, that the purification of matter and that of the spirit were a joint process, and that the same rules applied to what is above and to what is below, inspired some of the greatest minds of all time (and, inevitably, a long list of crooks) to study the nature and composition of matter and its many transformations. Every modern physicist who seeks to expand our knowledge of the fundamental properties of matter and our relation with the cosmos is following on a pathway alchemists treaded long ago.

CHAPTER 18

ADMIRABLE FORCE AND
EFFICACY OF ART AND NATURE

*(Wherein we visit the world of alchemy, an exploration
of powers hidden in matter through method and
spiritual discipline)*

The transformative powers of Nature are obvious for all to witness. That the heating, cooling, and mixing of the elements leads both to new compounds and to the reemergence of pure substances did not escape people's attention since at least ancient Egypt, probably much earlier. Could such natural powers be harnessed and used to extract the essence of things? Alchemy, at its most fundamental, was an attempt to re-create the powers of Nature and accelerate its transformations through experimental practices involving what would later become the key ingredients of chemical analysis: distillation, sublimation, the mixing of different chemicals and compounds—the collection of laboratory techniques that alchemists solemnly referred to as the "Art."

The widespread association of alchemy with dark magic and esoteric religion is, in many ways, a distortion of its true objective: the perfecting of metals and of the human spirit. Although Eastern, Jewish, Muslim, and Christian alchemists each combined their specific religious beliefs with their search for meaning, and as inspiration to achieve their goals, the unity of their practice is found in the common belief that a person could, in the confines of a laboratory, explore natural powers to effect material transformations. Implicit (and sometimes explicit) to this power was the approximation of human to deity: the successful alchemist transcended the mere human to become godlike—if not godlike, at least conversant with the ways of the Creator. Many alchemists believed that the "elixir," the compound able to purify metals into gold, would also extend life by bringing immunity to disease and by stopping aging.[1]

There isn't a fixed set of alchemical practices, as the body of knowledge changed in time and from place to place. Nevertheless, fire, with its power to transform matter, was always the key agent of change. If Earth's hot innards could churn matter into compounds or separate them into pure or almost pure metals, a person also could, through diligence and method, exact similar transformations in their furnace. Potentially, the practitioner could go even further, finishing Nature's incomplete work of transmuting all metals into the most perfect of them all—gold. In the words of the remarkable fourteenth-century Franciscan friar, alchemist, and natural philosopher Roger Bacon, "I must tell you that Nature always intended and strived to the perfection of gold: but many accidents coming between, change the metals."[2]

As fire was tamed and used for cooking, for warmth, and to scare away predators, soon it was seen to possess other less obvious properties, such as the power to transform certain minerals into metals. Fire appeared to be a magic knife, capable of extracting a purer essence than eyes could see, of carving out secrets hidden deep within matter.

As early as five thousand years ago, the bright green mineral malachite was being systematically burned in many sites in the Middle East to extract copper. Its discovery dates from probably twice as long, not much later than the appearance of the first agrarian communities. To watch the metal "escape" from the burning mineral must have conjured images of it being imprisoned within it:

fire, the destroyer, became also a liberator of matter's inner essence. As copper has a relatively low melting point of 1,981.4 degrees Fahrenheit (1,083 degrees Celsius), artisans were able to shape it into cups, jewelry, plows, and various other practical utensils. Weapons, though, necessitated something harder.

The need for military supremacy was surely a driving force in the search for more enduring metals capable of withstanding blows and of keeping their shape when sharpened, as Jared Diamond explored in his *Guns, Germs, and Steel*.[3] Then and now, the most advanced technology usually wins wars. The first answer was bronze, a mixture (amalgam) of copper and the even softer tin, usually at an 88 to 12 percent proportion, although there are many variations. How bronze was discovered is not known, but probably by trial and error. That two soft metals could combine to make something harder must have appeared to be quite a mystery.[4] By 3000 BCE, the early Bronze Age, different regions in the Middle East had bronze artifacts and weapons. In China, bronze art reached unprecedented beauty and sophistication, especially around 1500 BCE, during the Chang dynasty. By then, fire had become humans' greatest ally, a tool to explore Nature's hidden powers of transformation.

The dangers of this alliance are represented in many narratives, but none is more evocative than the Greek myth of Prometheus, the Titan who forged humans out of clay. In a daring act, Prometheus stole fire from the gods and gave it to humans. Enraged, Zeus had immortal Prometheus chained to a rock. That being not enough, Zeus had an eagle devour Prometheus's liver. As the liver of the immortal Titan regenerated each day, the torment went on and on to no end. If for a human martyr suffering is dreadful, for an immortal martyr suffering is dreadful and endless. Fire had to be quite a secret for poor Prometheus to deserve such a fate. Indeed, having control over fire was an encroachment into the gods' territory, something Zeus would not tolerate. Similar morals are echoed in Adam and Eve's fall from Eden, in Genesis 3; eating from the tree of knowledge of good and evil led to the expulsion of Adam and Eve from Eden and rendered them mortal. Although details vary across different faiths, the overall theme is the same: too much knowledge of Nature's innate powers is dangerous.

Human exploration of metals moved from bronze to iron, and from the Bronze to the Iron Age. Early iron samples were first recovered from meteorites, which often have a large amount of iron and nickel. Having a melting point about 500 degrees above copper, iron is harder to smelt but easier to find. By 1300 BCE iron smelting and smithing techniques were practiced in Anatolia (Turkey), in India, and also in the Balkans and the Caucasus region. As tin became harder to find, iron took over. Soon the addition of a bit of carbon to iron (usually less than 2 percent) was found to lead to steel, the hardest of the metal amalgams.

The development of the various amalgams reveals the germs of a scientific methodology: positive results were only possible from a detailed exploration of the properties of the different metals and their mixtures, always with the help of fire. There was also the understanding that the same practices led to the same results—that is, that there was regularity in Nature. Even if a premeditated search for the natural causes behind the material transformations in furnace and forge was lacking, an understanding was clearly emerging that using natural powers, humans could manipulate matter to serve their own needs. This knowledge was often considered sacred: those who had it were closer to the divine. Alchemy was born from the marriage of the practical and the holy, from the expectation that knowledge of Nature's secret powers brought one closer to divine wisdom.

Of the three main strands of alchemy, Chinese, Indian, and Western, we are mostly concerned with the last. There is no need to review the fascinating early history of alchemy here, only to call attention to its relation to the corpuscular theory of matter and to its pivotal role in the emergence of modern science. A key player in this connection was the late-eighth-century Jabir ibn Hayyan, court alchemist of the Abbasid caliph, Harin al Rashid, in Baghdad, then the center of the Muslim world. Known also by his Latin name Geber, Jabir allegedly was the first to use crystallization as a purification process and to isolate several acids: citric, tartaric, acetic, hydrochloric, and nitric. He may have combined the last two to create *aqua regia*, or royal water, a highly corrosive mixture that got its name from being capable of dissolving the royal metals gold and platinum.[5]

Jabir's distinctive mark was his attention to detail and methodology, both signatures of a nascent scientific approach. "The first essential in chemistry is that thou shouldest perform practical work and conduct experiments, for he who performs not practical work nor makes experiments will never attain to the least degree of mastery," he wrote.[6] Although his style, like that of most alchemists, is filled with bizarre symbolism and arcane imagery (the word "gibberish" may have originated from Geber), Jabir is credited with having used (and possibly invented) much of the equipment that became standard in chemistry laboratories, such as the alembic and a variety of retorts used for distillation. His vast opus—highly influential on alchemists in the Middle Ages—included the text from the fabled *Emerald Tablet*, the mysterious and presumably ancient work on alchemy attributed to legendary Hermes Trismegistus (Hermes the Thrice-Greatest), whose name is a syncretic mix of the Egyptian god Thoth, the god of wisdom and patron of the sciences, and the Greek god Hermes, the messenger.

The *Emerald Tablet* is important not only as the main sacred document of alchemists but for posing the unity of the cosmos as the fundamental principle of alchemy: "on Earth as in heaven." It consists of thirteen mysterious one-liners that supposedly hid the key to the secrets and goals of alchemy. A translation was found among Newton's vast alchemical writings, a stunning proof of its influence on one of science's central figures. In particular, in Newton's translation the second entry reads: "That which is below is like that which is above, that which is above is like that which is below to do the miracles of one only thing."[7] Newton's theory of gravity, describing the attractions of masses on Earth and in the sky as having the same nature, was a concrete expression of this alchemical principle, as I noted in Part 1. The unification works both ways: the heavens are brought closer to Earth, and Earth is brought closer to heaven; those who understand this are closer to God's mind. Recent work on Newton's life and work from Betty Jo Teeter Dobbs and other historians of science leave little doubt that this was Newton's main motivation.

The central goal of alchemy, as many know, is the transmutation of "impure" metals into the purest of them all: gold, the metal that doesn't rust.[8] "Alchimy therefore is a science teaching how to make and compound a certain medicine, which is called *Elixir*, which when it is cast upon metals or imperfect bodies, doth fully perfect them in the very projection," wrote Roger Bacon in *The Mirror of Alchimy*. The "philosopher's stone," or elixir (a word derivative from the Arabic *al-iksir*, "the effective recipe"), was the active catalyst able to remove the impurities and to complete Nature's interrupted work. According to Bacon, following the prescription Jabir had set up, the two "principles" within different metals were mercury and sulfur, in different amounts: "For according to the purity and impurity of the aforementioned principles, mercury and sulfur, pure and impure metals are engendered: to wit, gold, silver, steel, lead, copper, and iron."[9] So sulfur is the pollutant, inflammable and transitory; mercury the cleanser, dense and permanent. From the relative amounts of the two a greater or lesser degree of purity is achieved.

In some traditions, the Elixir was also capable of affecting the alchemist; the Philosopher's Stone purified metals and souls, lifting humans from the greatest burdens of all: illness and mortality. The laboratory process of purification, through its arduous repetition and devotion, cleansed the human soul; only those of true purity could ever expect to be successful in their quest.

Considering science as a body of knowledge resulting from methodically studying the workings of Nature, we can see how alchemists—the scrupulous and the unscrupulous—intended to use the science of their time to lift humankind, or at least themselves and their patrons, from the burdens of poverty and illness. The German-Swiss physician and alchemist Paracelsus, active during the early sixteenth century and founder of toxicology, stands out as a perfect example of the bridging between occult and scientific practices. We identify here a trend that remains very much alive today in scientific research: the creation of wealth and medications through the scrupulous and unscrupulous manipulation of Nature's resources. If one of science's goals is the alleviation of human suffering, its roots reach back to ancient alchemical practices.

In Aristotle's description of natural transformations, each of the four basic elements had different qualities that could be exchanged. Earth was dry and cold, water was moist and cold, air was moist and hot, and fire was dry and hot. Material transformations unfolded through the exchange of the various qualities and the mixing of the elements. According to historian of science William R. Newman, Jabir adapted Aristotle's notions of dryness and moistness to the two essential elements, sulfur (dry) and mercury (moist). Alchemical practice thus endeavored to change the relative amounts of these qualities, which occurred in different proportions for the various metals. When Pseudo-Gerber wrote the influential *Summa Perfectionis* (The Sum of Perfection) in the thirteenth century, he promoted Jabir's qualities to properties belonging to corpuscles of sulfur and mercury, which could have different size, purity, and relative fractions. Following the Greek Atomistic tradition, the corpuscles didn't change, retaining their essence through the various chemical processes, even if composed of the even smaller particles of the four basic elements: "Hence the mercury and the sulfur themselves form secondary particles of larger size than their own elementary components, and these secondary particles, because of their strong composition, have a semi-permanent existence," noted

Newman in his study of Pseudo-Gerber's writings.[10] The similarity with the modern notion that a few fundamental particles (electrons, protons, and neutrons) compose the atoms of different elements is striking. Or, for that matter, of atoms composing molecules.

Such concepts present in Pseudo-Geber writings are strong evidence in favor of a pervasive corpuscular interpretation in alchemy, which influenced none other than Robert Boyle, the seventeenth-century natural philosopher considered by many to be the father of modern chemistry, and from whom Isaac Newton apprenticed his alchemy. Science had not yet distinguished itself sharply from its progenitors. Boyle's mechanical philosophy, in which he strived to show matter as being composed of particles having only properties of size, shape, motion, and texture, had its origins in late medieval alchemy.

Newton went to Boyle for alchemical secrets, but Boyle apparently told him little. One coveted substance Boyle had synthesized, known as "red earth," was believed to be only one step removed from the Philosopher's Stone: able to convert lead into gold, but not as efficiently. Newton would behold a sample of red earth only after Boyle's death in 1691, thanks to Boyle's executor, the empiricist philosopher (and alchemist) John Locke. Another coveted substance was philosophical mercury, a liquid form of mercury able to dissolve gold slowly, and an important step toward the final goal. Lawrence Principe, a chemist and historian of science from Johns Hopkins University, has followed the recipes of Boyle and others and managed to make philosophical mercury, after much trial and error. In the best alchemical tradition, Principe mixed philosophical mercury with gold and sealed it in a glass egg. As he told science journalist Jane Bosveld, "the mixture began to bubble, rising 'like leavened dough.' Then it turned pasty and liquid and, after several days of heating, transformed into a 'dendritic fractal'; a metallic tree, like the trees (or metal veins) miners see underground, only this one was made of gold and mercury."[11]

The notion that metals grow underground as branches in trees provided an organic dimension to the alchemy Newton and Boyle practiced. The laboratory was a facilitator, a place where the alchemist could reproduce Nature's work, perhaps succeeding in accelerating it through diligent methodology. And even if Newton kept secret his one million plus written words on alchemy, using a cyphered code that remains to be broken, some of his alchemical viewpoint—both the organismic and the Atomistic—leaks into his "exclusively" scientific writings, such as the *Principia* or *Opticks*. At the end of Book III in the *Principia* Newton writes,

> And the vapors that arise from the sun and the fixed stars and the tails of comets can fall by their gravity into the atmospheres of the planets and there be condensed and converted into water and humid spirits, and then—by a slow heat—be transformed gradually into salts, sulphurs, tinctures, slime, mud, clay, sand, stones, coral, and other earthy substances.[12]

It's easy to identify in this passage the alchemist's belief of Nature's working (by slow heat) to create all sorts of compounds from a primal cosmic substance. Still in the *Principia*, in the Introduction, Newton expresses his belief in an atomistic description of matter: "For many things lead me to have a suspicion that all phenomena may depend on certain forces by which particles of bodies, by causes not yet known, either are impelled toward one another and cohere in regular figures, or are repelled from one another and recede."[13]

Newton's belief that "many things lead me to have a suspicion" probably came from his alchemical experiments; same with "all phenomena may depend on certain forces," expressing his belief in the unity of Nature, where a few forces explain a myriad of phenomena, and, finally, his notion of attraction and repulsion of the "particles of bodies," that "cohere in regular figures," a prescient foray into how larger atoms form and even how atoms combine to make molecules with certain symmetries.

But it is in his *Opticks* that Newton would let go, so to speak, and speculate more freely about the nature of matter and light, often with unbelievably accurate intuition: "Do not all fix'd Bodies, when heated beyond a certain degree, emit Light and shine; and is not this Emission perform'd by the vibrating motions of their parts?"[14] This is precisely what happens, as electromagnetic radiation (sometimes in the form of visible light) is emitted from heated bodies resulting from the internal vibrations of their solid structure and to electrons jumping between different atomic orbits. (More on this later.)

Light, too, was seen as composed of corpuscles: "All Bodies seem to be composed of hard particles: For otherwise Fluids would not congeal. . . . Even Rays of Light seem to be hard bodies . . . And therefore Hardness may be reckon'd the Property of all uncompounded Matter."[15] Quite remarkably, especially when we consider what we now know about the properties of light, Newton even speculates on the possible transmutation of light into matter and vice versa: "And among such various and strange Transmutations, why not Nature change Bodies into Light, and Light into Bodies?"[16] That this very transformation is at the core of Einstein's theory of relativity (embodied in the $E = mc^2$ formula) is nothing short of marvelous.

CHAPTER 19

THE ELUSIVE NATURE OF HEAT

*(Wherein we explore phlogiston and caloric, the strange
substances proposed to explain the nature of heat, and how
such substances were later discarded as explanations)*

From so murky a beginning, infused in mercury vapor billowing from the alchemist's crucible, inspired by visions of celestial perfection, science took a sharp turn toward a new era in which such dreamlike musings of the past were increasingly seen as an embarrassment. No more God talk in scientific treatises, no more spiritual parlance in dealings with natural phenomena: only precise, mechanistic rhetoric would do, dressed in the robes of mathematical rigor. Newton's theory of Nature, based on how material objects large and small reacted to attractive and repulsive forces between them, became the beacon of the Enlightenment. The world, in all of its complexity, could be methodically studied by breaking it down to its simplest parts, each following the path determined by the sum of the forces acting upon it. Newtonian physics catapulted the triumphal ascension of reductionism.

The pace of change picked up. If forces kept bits of matter together, they had to be overwhelmed to release their grip. As in alchemy before, heat was the key ingredient. When heated, ice changed to water, and water changed to vapor. Most substances would respond to heat in one way or another. Gases would expand, increasing their pressure; solids—even hard metals—would melt, becoming liquid. Already in 1662 Robert Boyle had demonstrated that the pressure and volume of a given amount of gas at a fixed temperature are inversely proportional: if you trap gas in a container and squeeze it down with a piston, the gas pressure rises in the same proportion that the volume decreases. The trio of macroscopic variables—pressure, volume, and temperature—being directly measurable, would allow a quantitative study of gases and their properties: if the volume of a gas is kept fixed and the temperature is increased, its pressure increases in the same proportion. If, instead, the pressure is kept fixed while the temperature is increased, the volume of the gas increases in the same proportion.[1]

The remarkable point is that this same proportionate behavior is true for *any* gas. This is how laws emerge in physics: a regular trend is identified in a few examples and then generalized to apply to a whole class of substances or entities. The generalization is tested as broadly as possible, until it starts to fail. In the case of gases, it may fail because extreme conditions are eventually achieved, leading to a change of circumstances and the breakdown of the law. For example, very high pressures may liquefy or even solidify a gas. Surely, this general behavior and its breakdown must be a consequence of whatever makes up gases?

The answer was found and forgotten during the early eighteenth century, then found again one hundred years later and summarily rejected, before finally being picked up decades later, not without controversy. The stubborn rejection and controversy were not completely baseless. The proposed answer made possible a dangerous precedent, that physical explanations could claim the reality of invisible realms existing beyond our senses or even the sensitivity of measuring devices. Could something we can't see or even be sure to exist explain what we measure? If so, where do we draw the line between an invisible reality inaccessible to our devices and a fantastical hypothesis? To put it bluntly, if atoms and fairies are both invisible, why do we claim that atoms exists and fairies don't?

   In 1738, the brilliant Dutch mathematician Daniel Bernouilli proposed, true to the Atomistic worldview, that gases consist of a vast number of tiny molecules in random motion. These molecules would move about and bounce against one another without losing much energy in the process. Based on this Atomistic hypothesis, Bernouilli showed that the pressure of a gas is due to the collisions of the molecules with the walls of the container: Boyle's law states that if the volume containing the gas is halved while its temperature is kept constant, its pressure doubles. Based on his microscopic hypothesis, Bernouilli argued that as the volume of the gas is decreased, molecules are being squeezed, and their collisions against the container become more frequent; the macroscopic effect of the countless collisions is a net increase in the gas pressure. In other words, Bernouilli attempted to explain a macroscopic property of a gas, its pressure, in terms of invisibly small, microscopic entities. Would Atomism finally become a quantitative science?

   Nothing much happened until 1845, when the British physicist John James Waterston submitted a paper to the Royal Society where he claimed to relate the temperature and the pressure of a gas to its tiny molecular constituents. He showed that the temperature of a gas is proportional to the square of the average velocity of its molecules and that its pressure is proportional to the density of the molecules times the square of their average velocity.[2] This was an unprecedented attempt to relate temperature with motion— more shockingly, with motion of invisible entities.

   For centuries, scientists had struggled with the nature of heat, inventing fanciful explanations to make ends meet, confusing heat with combustion. First came the phlogiston, a somewhat magical substance that allowed something to burn, proposed as early as 1667 by the German alchemist and physician Johann Joachim Becher: flames appeared when combustible substances released phlogiston. As a consequence, a "dephlogisticated" substance would not burn. The phlogiston hypothesis was challenged when metals were shown to gain weight when burned. Wild speculation ensued, with some proposing that phlogiston had *negative* weight, while others that it was lighter than air. This is not an uncommon trend in science: when a compelling idea starts to sputter, progressively more manic attempts to rescue it are put forward, even if seemingly requiring desperate measures. Only in 1783, when Antoine-Laurent Lavoisier demonstrated in a series of revolutionary experiments that burning requires a gas with weight (oxygen), and that in every chemical reaction—including burning—the total mass of the reactants remains constant, was the phlogiston idea finally abandoned.

Having explained combustion, but still mystified by the nature of heat, Lavoisier proposed the existence of a new substance, the caloric. He suggested that the flow of heat from hot to cold was due to the flow of caloric. Given that the total mass in any reaction is constant, Lavoisier further suggested that caloric was massless and that its total quantity was conserved in the Universe. Many explanations of heat-related phenomena ensued, all wrong in spite of their apparent reasonableness. For example, hot tea cooled off because caloric, which has higher densities in hotter areas and is self-repellent, would thus slowly flow from hotter to colder areas (from the hot liquid to the cooler air around the tea cup). The caloric was a sort of aether endowed with flowing ability, weightless yet effective in explaining many natural phenomena.

The first challenge to the caloric hypothesis came from Count Rumford, a Loyalist from New Hampshire with a life history worthy of a major epic movie. In one of his many employments after leaving the United States, Rumford worked as a munitions expert in Bavaria, in particular overseeing the boring of cannons. As a huge drill bored a hole in a cylindrical piece of metal, water was used to cool off the tremendous heat released by friction. Rumford noted that as long as the drilling persisted, heat never ceased to leak from the metal, and the water kept on boiling. In 1798, he wrote in his observations, "Anything which an insulated body, or system of bodies, can continue to furnish without limitation, cannot possibly be a material substance."[3] He went on to suggest that it is not the transfer of caloric that caused the flow of heat but the friction between drill and metal. Heat, he conjectured, is matter in motion, not a substance. Although the scientific community did not embrace Rumford's claims immediately, his experiment planted the seed that perhaps heat is not a substance but a property of substances.

The second, and more deadly, challenge to the caloric hypothesis came from James Prescott Joule, who devised a series of detailed experiments in the 1840s to determine how mechanical work can lead to an increase in temperature. Joule used rotating paddles to agitate water in a large vat to make the determination precise, relating the amount of required mechanical work to raise the water temperature by 1 degree Fahrenheit. Joule's results were an expression of energy conservation and transfer, the proper way to describe how substances get heated or cooled. As the paddles churn the water about, the water molecules gain energy, and their average speed increases; this increase in speed is related to an increase in temperature, just as Waterston had proposed. Joule was aware of the work of both John Herapath (see note 2) and James Waterston on the microscopic theory of gases. He was a pupil of none other than John Dalton, the champion of atomic theory, who early in the nineteenth century had proposed that chemical reactions amounted to precise exchanges of atoms among compounds. For example, tin could combine with one or with two atoms of oxygen, and the masses of the resulting compounds reflected the number of oxygen atoms in each. Dalton proposed that each chemical element had its atom, and that chemical reactions were unable to break them down. Furthermore, atoms of different elements could combine to make all sorts of chemical compounds, which we nowadays call molecules.

Between the microscopic theory of gases and Dalton's Atomistic explanation of chemical reactions, the notion that matter had a corpuscular substructure slowly gained weight. The rise and fall of the phlogiston and the caloric as explanations for combustion and heat illustrate quite powerfully the workings of science. As scientists strive to explain natural phenomena, they will conjure new hypotheses, which they will arduously defend. This is precisely how it should be, given that the more compelling the idea, the more passion it incites in its proposers and followers. However, since scientific hypotheses must undergo constant empirical scrutiny, they will suffice only until proven wrong or limited in their scope. An explanation may seem satisfactory to describe the data (to "save the phenomena," as Plato would have put it) even if ultimately wrong. Epicycles described the celestial motions with good precision, even if completely artificial; phlogiston and, more so, caloric described combustion and heat flow, even if they were completely unphysical. The power of science to narrow in on ever more accurate descriptions of physical reality relies fundamentally on our ability to test hypotheses with ever greater precision. If this drive toward greater precision is blocked or interrupted, scientific advance stagnates. Research pushes the boundaries of the Island of Knowledge outwards (and sometimes back in). That there are no lighthouses out in the ocean of the unknown to light the way adds to the challenge and the excitement of the scientific quest. Few examples illustrate this quest as clearly as the study of light and its elusive nature.

CHAPTER 20

MYSTERIOUS LIGHT

*(Wherein we explore how light's mysterious properties spawned
the twin scientific revolutions of the early twentieth century)*

W e are creatures of light—this elusive, bizarre entity that even
today mystifies most of us.

Light coming from the Sun is the sum total of many electromagnetic waves, each with its own wavelength. The small window
that we see, the visible portion of the spectrum from violet to red, is
made of waves with wavelengths between 400 and 650 billionths of
a meter (nanometers). A wavelength is simply the distance between
two successive wave crests. So when we speak of short wavelengths,
we mean wave crests packed close together; longer wavelengths'
wave crests are more spread out.

We are very much a product of four billion years of evolution
on a planet bathed in bright sunlight. With a surface temperature
of 5,505 degrees Celsius (9,941 degrees Fahrenheit), the Sun is in-
formally classified as a yellow dwarf star, emitting most intensely in
the yellow-green spectrum. Although the Sun's surface is white, the
yellowish color we perceive here on Earth comes from the scattering of blue frequencies as sunlight passes through the atmosphere.
The brightness we see during daytime comes from light bouncing
about the nitrogen and oxygen molecules in air. Dust helps. This
bouncing also explains the blueness of the sky during daytime: air

scatters short wavelengths more efficiently than long ones, and blue has shorter wavelength than yellow or red. So if we look at the sky away from the Sun, we predominantly see the portion of sunlight that is most efficiently spread about, comprising blue mixed with some white.[1] Given that air molecules are thousands of times smaller than the typical light wavelength, we can see why blue light gets spread the most. Like a huge wave passing over a small rock, yellow or red light, being a longer wavelength, can hardly notice the small molecular obstacles in its way. At sunset, sunlight hits Earth tangentially and travels longer through the atmosphere, and most of the blue tones scatter away before reaching low altitudes. As a result, we see more red and orange than blue and green. During overcast days, the water droplets and frozen crystals that make up clouds scatter all wavelengths of sunlight somewhat uniformly, and the result is a whitish glow.

Contrary to what we naively may think, the light our eyes see is less than half of the total radiation the Sun sends our way. Without scientific tools to reveal what is invisible to the eye, we would have a very limited knowledge of physical reality. Even so, it is important to keep in mind that from within our Island of Knowledge the finite reach of our instruments always limits our improved vistas. And as we see more, the more we know there is to see.

Visible light constitutes only about 40 percent of the total radiation output of the Sun as it hits the top of the atmosphere; 50 percent infrared and 10 percent ultraviolet constitute the rest. Down where we are at the surface, thanks to the atmosphere's protective work, only 3 percent of ultraviolet remains, while visible light goes up to 44 percent of the total. When it comes to the Sun (and so much else), what you see is not all that you get. Our sensorial perception has been naturally selected to maximize our chances of survival on this planet. On another planet, with more or less starlight and different atmospheric composition and thickness, creatures would evolve to sense other parts of the electromagnetic spectrum. Even here, night-dwelling creatures or those living in dark caves or deeply underwater have adapted in different ways: think of bats using echolocation and the bioluminescence of deep-sea fish.

The explanation of the color of the sky and related results represent a triumph of late-nineteenth-century physics, when light was described as vibrating electromagnetic fields. Every source of electromagnetic radiation can be traced back to electric charges oscillating or accelerating. Between 1861 and 1862, Scottish physicist James Clerk Maxwell was working at my alma mater, King's College London, when he obtained a relationship between electricity and magnetism that led to a new way of portraying how objects interact with one another. Up to then, interactions were described in terms of forces, as in Newton's gravitational force, or the force we apply on a bike when we push it uphill. Inspired by ideas from Michael Faraday, Maxwell proposed his celebrated theory of the electromagnetic *field*. Since then, the concept of field became the way physicists describe how objects as varied as stars and electrons interact: a force is derived from a field.

The concept became so powerful that it migrated away from interactions: we can talk of the field of temperatures in a room (how the temperature varies from point to point) or the velocity field of the water in a river or the wind in the atmosphere. An electric charge creates an electric field around it; this field is a spatial manifestation of the charge, so that another charge approaching it would "feel" its presence before touching it; the closer to the charge, the stronger its field. Same charges repel; opposite charges attract. The same happens with magnets, as a quick experiment can demonstrate: bring two fridge magnets close together, and they will repel, resisting further intimacy. The space around the magnets seems to be filled with something that tells the magnets to repel. This "something" is the magnetic field the two magnets create.

Likewise, your mass creates a gravitational field around you; other masses feel your field and are attracted to it in inverse proportion to the square of their distance to you.

When an electric charge oscillates, its electric field oscillates with it. As a more familiar visual, a cork bobbing on water emits (two-dimensional) waves that propagate outwards on the surface of the water as concentric circles. Likewise, oscillating charges emit electric waves but in three dimensions. As it accelerates up and down, it also creates a magnetic field, which oscillates along with the electric field. One field bootstraps the other, as they both move in tandem away from the charge. The only difference is that they point perpendicularly to each other like the two directions of a cross: if the charge is bobbing up and down, the magnetic field bounces left and right, and the wave travels in the direction perpendicular to the cross (we say that electromagnetic waves are *transverse*).[2]

To summarize, charges moving about create waving electric and magnetic fields that propagate through space. Maxwell showed that in empty space the speed of such propagation is the speed of light. Such a result led him to an amazing conclusion: light is electromagnetic radiation, electric and magnetic fields propagating as waves. The only difference between, say, red light and ultraviolet is that red light has longer wavelength than ultraviolet. From long to short wavelength, the electromagnetic spectrum contains several types of radiation: radio, microwave, infrared, visible, ultraviolet, x-rays, and gamma rays, the shortest and most energetic of the bunch.

If light (as mentioned earlier, I'm using "light" to represent all kinds of electromagnetic radiation) is a wave, where does it wave? After all, more familiar waves are vibrations of a medium: water waves are vibrating patterns on water; sound waves are pressure waves on air; if you grab one end of a rope and shake it, the waves are happening on the rope. So where do light waves happen? This is the first of the many mysteries associated with light. We now know that light doesn't need any material medium to propagate: it propagates in empty space, the bootstrapping action of the electric and magnetic fields being all it needs to push itself forward. Of course, light can propagate through material media as well, as we know when we swim underwater or look through glass. The net effect of having light move through different media is to slow it down, as the light waves will cause the electric charges making up the medium to bob along as they pass by.

It was clear to nineteenth-century physicists that light was different, since there was no obvious material medium to support its propagation. Still, they believed, *something* had to do the job. Maxwell, for one, spent many precious years devising increasingly bizarre mechanical models to make sense of electromagnetic wave propagation through space. A new kind of medium—the luminiferous aether—was invoked solely to provide support for light waves. Two centuries earlier, Newton and the Dutch Christiaan Huygens had independently pondered the nature of light and arrived at opposing views. Newton, ever the Atomist, proposed that light was made of small corpuscles but struggled to show that all of its properties were consistent with this assumption: an easy task for transmission and reflection of light (through straight lines), but harder for refraction—the change in direction of propagation when light moves between different media—and diffraction—the spreading out of waves when passing through a narrow obstacle. Huygens, in contrast, proposed that light was a wave propagating in an aether-like medium.

The ping-pong between particle and wave continued until early in the nineteenth century, when Thomas Young and Augustin-Jean Fresnel independently argued for light as a form of transverse wave. Young, in particular, devised a series of experiments involving diffraction that conclusively demonstrated that light was a wave. Cutting a rectangular hole through a paper card, Young stretched a human hair across the hole and illuminated it with a candle. As he related in a document from 1802, "When the hair approached so near to the direction of the margin of a candle that the inflected light was sufficiently copious to produce a sensible effect, the fringes [alternating dark and bright stripes] began to appear; and it was easy to estimate the proportion of their breadth to the apparent breadth of the hair across the image of which they extended."[3] By the time Maxwell had shown that light was a transverse electromagnetic wave, Newton's corpuscular theory had been pushed out of the limelight. Similar interference patterns are easily seen as water waves pass through obstacles. Light seemed to be doing the same.

However, the more the nature of light was elucidated, the stranger the aether proved to be. As with the phlogiston and the caloric, "magical" was probably a better word. The luminiferous aether had to be a fluid to fill all of space, somewhat like the Aristotelian aether of ages past; however, it also had to be millions of times more rigid than steel—to support the propagation of short wavelengths—and transparent, so that we could see light from distant stars. Mysteriously, it also had to be massless and without any viscosity, so as not to interfere with planetary orbits. That the vast majority of the brightest scientists of the time embraced such an odd entity with absolute confidence illustrates how hard it is to shake off prejudice born out of experience: a wave had to wave on something. To the scientists of the late nineteenth century, the aether was easier to consider than the possibility that light could simply travel in empty space. Once again, the cosmos seemed to be embedded in some sort of dilute substance, inaccessible to the senses.

To be a viable physical entity the aether had to be detected, directly or indirectly. Given its otherworldly properties, direct detection was out: to detect something requires it to interact with the detector. And what kind of device can detect what is imponderable and with no viscosity? The existence of the aether had to be confirmed through indirect effects. Not an easy challenge.

In 1887, Albert Michelson and Edward Morley conducted a brilliant experiment to measure the aether's effect on the propagation of light. The idea relied on the assumption that if the aether existed, it had to be an inert medium at absolute rest. Think of the air on a perfectly still day. Maxwell had shown that electromagnetic waves propagated at the speed of light with respect to the stationary aether. But since Galileo's days, it was known that speeds are usually measured relative to a given reference. For example, if you are standing in front of a shop and a car drives by, you measure its speed with respect to your state of rest. If, instead, you were biking along the same direction as the car, the car's speed would be smaller relative to you. The notion of an absolute frame of reference violated this notion of relativity, since it meant that all speeds could be measured with respect to the aether. Shocking as this possibility was, the alternative—light traveling in empty space—was considered worse.

So Michelson and Morley had a clever idea. Since the Earth is moving about the Sun, it should feel an "aether wind" against its direction of motion. The same happens when you bike or drive on a perfectly windless day. You feel the air blowing against you as you move. If a ray of light were shot against the aether wind, its propagation speed would be slower than if the Earth weren't moving against the aether. If, otherwise, it were shot at a right angle with the direction of the motion, it shouldn't suffer any delay. It was then a great shock when Michelson and Morley performed the measurement in two perpendicular directions and declared that they couldn't detect any difference at all; their experiment indicated that light traveled with the same speed in all directions. If there were an aether, light was clearly indifferent to it, contradicting the aether's raison d'être.[4]

Panic ensued. Many physicists tried to come up with plausible explanations as to why the experiment had "failed." Most notably, the Irish physicist George FitzGerald and the Dutch Hendrik Antoon Lorentz independently proposed that any material object that moved against the aether would shrink a bit, including the measuring apparatus; the faster the motion, the greater the shrinking. The shrinking, if real, would explain why there was no difference in the experiment: light did slow down as it moved against the aether wind, but it also had to travel a shorter distance because the arms of the measuring apparatus shrunk. The two effects neatly cancelled out so that the net result was just what Michelson and Morley had measured.

Although some scientists were relieved, no one was truly convinced, since the space contraction hypothesis seemed to have come from nowhere. And even if FitzGerald and Lorentz were right, there was still the very basic question as to why, contrary to all of Newtonian physics, in which the laws of Nature were the same for any nonaccelerating frame, electromagnetism seemed to require a universal reference frame. The pillars of the classical worldview, Newtonian mechanics and Maxwell's electromagnetism, were deeply at odds with one another. Something was profoundly wrong. But rescue was soon to come.

Einstein starts his famous 1905 paper on the special theory of relativity expressing his concern that Maxwell's theory seems to require an absolute reference frame. He then points out that results in electromagnetism and all of physics should be the same for every observer moving at any constant velocity and that "unsuccessful attempts to discover any motion of the earth relatively to the 'light medium,' suggest that the phenomena of electrodynamics as well as of mechanics possess no properties corresponding to the idea of absolute rest."[5] His revolutionary paper incorporates the notion that space does shrink in the direction of the motion and that moving clocks or, more generally, time slows down. So the length contraction FitzGerald and Lorentz had proposed wasn't incorrect. What was incorrect was their interpretation of it, which presupposed the existence of an inert universal medium. Einstein did away with the aether, explaining that Maxwell's electromagnetism is perfectly consistent in any inertial reference frame (i.e., moving at constant velocity) *so long as something else is imposed*, a new postulate, as Einstein called it: "light is always propagated in empty space with a definite velocity c which is independent of the state of motion of the emitting body."[6]

The trade-off for the aether not being an absolute reference frame (or not existing) was that light always travels with the same speed. Einstein substituted one absolute for another! He had no proof that he was right; his guiding principle was that the laws of physics should be the same in every inertial reference frame, that is, that Nature should exhibit this fundamental symmetry. How could science make sense if every observer had a different set of laws and measurements? He thus elevated the principle of relativity (that the laws of Nature are the same in all inertial frames) to the level of a postulate, assuming it to be true.

Even so, his second postulate was by far the most daring. Why should light be different from everything else? What makes it always have the same speed? Einstein didn't know why light had a constant speed or why its value was 299,792,458 meters per second (186,282 miles per second). He assumed this to be true in order to reconcile electromagnetism with the principle of relativity. The constancy of the speed of light was the price that had to be paid to restore order to physics. In so doing, Einstein whisked the aether away by making light doubly mysterious: a wave that could propagate in emptiness and always with the same speed. And he was only getting started.

The paper on the special theory of relativity was one of four papers the then twenty-six-year-old Einstein published in 1905. The first paper he submitted that year was, by his own opinion, his most revolutionary. The title was inconspicuous enough: "On a Heuristic Point of View About the Creation and Conversion of Light." Einstein starts by making the point that Maxwell's wave theory of light was at odds with the prevailing notion that matter consists of atoms and electrons; whereas waves are continuous in space, atoms are discrete entities. He then puts forward his "heuristic point of view," that just like matter, light might be considered as being made of small little bits so that "incident light consists of energy quanta with an energy [h x f]."[7] h is Planck's constant, a constant of Nature associated with all quantum phenomena, and f is the frequency of the light beam. If light is not monochromatic (that is, if light is composed of waves with different frequencies) there will be many types of such quanta, one kind for each frequency. If Einstein were right, light would regain its particle-like interpretation. Newton would have rejoiced.

Einstein was careful to note that the wave nature of light was still valid, so long as it wasn't "applied to the phenomena of the creation and conversion of light."[8] In other words, light's granular behavior complements its wavelike behavior as the other side of the same coin. Light could be seen as either, depending on the nature of the physical phenomena under study. Likewise, water at room temperature is treated macroscopically as a fluid and microscopically as composed of individual molecules. What water is depends on the context; what light is depends on the context. In reality, light is neither particle nor wave.

The goal of physics is not to impose attributes on physical entities with any sort of finality (as in *this* is what water is or *this* is what light is) but to explain the result of experiments. The concepts scientists come up with are explanatory tools, devices created to give meaning to their measurements. To a physicist, what something "is" is less relevant than whether her explanations are efficient. In fact, as we descend into the strange world of the quantum, the meaning of "being" in the sense of permanent identity becomes untenable; nothing is what it seems, and nothing remains what it is for very long, as matter and light dance about in continuous transformation. With his heuristic idea, Einstein opened the door to the impermanence of the quantum world. It is only fitting that light illuminated the way in.

CHAPTER 21

⟨⟨⟨⟨⟩⟩⟩⟩

LEARNING TO LET GO

*(Wherein we begin our exploration of quantum physics and
how it imposes limits on what we can know of the world)*

W ithin a decade, Einstein's theories of special relativity and
of light quanta had turned physics upside down. From a
placid wave traveling in the luminiferous aether, light had become
profoundly mysterious: not just the fastest of all phenomena but
with a speed that, unlike anything else, was independent of the
motion of its source; a wave that, unlike any other wave, could
travel in nothingness; an entity that was *both* particle and wave,
thus defying commonsense intuition that things could only be one
or the other. Light's speed wasn't only the fastest known; it was
also *the* limiting speed, the maximum allowed in Nature. No signal
could travel faster than light; no information could arrive before
light did. Looking out to the near and far corners of the Universe,
physicists and astronomers learned that light *was* information, as
they carefully collected the many kinds of electromagnetic radiation that distant objects emit to build a picture of the world. And
in this book we have explored how this limiting speed leads to the
existence of a cosmic horizon beyond which we cannot receive any
information.

More mysteriously, light can only travel at its speed because
it has no mass. Those little grains of light, that later were called

photons, are massless bundles of pure energy. Physics was thus proposing that something could exist without mass, that things could exist without being material. Since what exists defines physical reality, the new physics suggested that reality could be immaterial. Energy is more fundamental than mass, more essential. A deeper understanding of Nature demanded a new worldview. Physicists had to let go of the old ways.

In his fourth paper of 1905, only a few pages long, Einstein derived the famous $E = mc^2$ formula. He wrote: "If a body gives off the energy L in the form of radiation, its mass diminishes by L/c^2."[1] Einstein went on to conclude that "the mass of a body is a measure of its energy-content." We could thus refer solely to the energy of things, in things. Energy unifies mass and radiation, making one potentially become the other. At the end of his paper Einstein speculated: "It is not impossible that with bodies where the energy-content varies to a high degree (e.g. with radium salts) the theory may be put successfully to the test."[2] How right he was! Indeed, the radium salts Einstein referred to are radioactive nuclei that emit either small particles or pure radiation as they decay. The radiation kind of decay, comprised of gamma-ray photons, has energy corresponding precisely to the mass loss of the nucleus (times the square of the speed of light), just as Einstein had predicted.

The next twenty-five years were nothing short of explosive. The quantum revolution was indeed a revolution, not only in the way we see the world but in the way we live in the world. Its rippling effects are still being felt today and will continue for a long time. Here we are mostly interested in the first aspect of the quantum revolution: its foundational impact on our conception of reality. The second aspect, the more pragmatic and technological implications of quantum physics, including the digital age we live in, is a related but separate matter. We will occasionally refer to the uses of digital technology in data acquisition and analysis, but as ancillary to our main goal.

The first fundamental lesson from quantum physics is that a commonsense view of the world based on our sensorial perception of reality, what we often call the "classical" worldview, is an approximation. Reality is quantum mechanical through and through, from bottom to top, from small to large. Classical descriptions such as Newton's laws of motion or Maxwell's electromagnetism work because quantum effects are negligibly small for large systems. We are quantum beings just as electrons are, but our quantum essence is extremely subtle—so subtle, in fact, as to be irrelevant for the most part. The same with trees, cars, frogs, and amoebae, although as we get to smaller and smaller objects, the sharp distinction between the classical and the quantum becomes increasingly fuzzy. The lesson is clear: as we descend into the realm of the very small, we must embrace a reality quite distinct from our own.

The first problem to solve was the structure of the atom. In 1911, Ernest Rutherford showed that an atom is made of a very massive and densely compact nucleus of positive electric charge surrounded by negatively charged electrons. The (wrong) image of an atom as a mini solar system is often invoked to illustrate Rutherford's results. The problem is that electric charges are not planets. Maxwell's electromagnetism explained that accelerated electric charges radiate away their energies. If that were the case, how would an electron orbit a nucleus for long periods of time without spiraling inwards kamikaze-style? Rutherford didn't know but was characteristically sure of his results.

In 1913, Danish physicist Niels Bohr offered an answer, albeit a very strange one: he proposed that electrons can only orbit the nucleus at fixed orbits, like steps on a staircase. Just as you can't balance between steps, the electron can't be between orbits. Each orbit has an energy associated with it: the higher the orbit, the higher the energy. In the staircase analogy, the higher the step, the more energy you need to jump to it. Conversely, going down releases energy.[3] Bohr's wild idea was to assume, without any prior reason, that once the electron reaches the lowest possible orbit (the lowest step of the staircase) it can't go down any further. The lowest orbit, called the "ground state," is a termination point.

Bohr didn't offer any reason as to why that was so. The strength of his argument was to combine classical notions of circular orbits with Planck's and Einstein's ideas of discrete (quantized) energies and light bundles to explain the types of radiation atoms emit when excited. Bohr suggested that for an electron to climb to a higher orbit, it needs to absorb an incoming photon with energy nearly equal to the energy difference between the two orbits. Just as we need energy to climb upstairs or to hike up a mountain, the electron "eats" the photon to climb up. In contrast, as it goes downward, it releases photons with energies equal to the difference between the two orbits. Since different atoms have different numbers of protons and electrons and thus different orbits (or energy levels), each has a unique emission spectrum: the sum total of the possible jumps the electrons can perform as they cascade down to the ground state. These spectral signatures, often compared to fingerprints because of their uniqueness, are the key ingredient of spectroscopy, the life-blood of astronomy. Instead of traveling to a distant star or galaxy to study its composition and properties, astronomers can simply collect and study its light and spectral signature.

Bohr's theory was clearly a hybrid, a transitional description. A more complete formulation of the behavior of electrons in atoms had to wait until the Great War was over and physicists could think about physics again. There were two main schools of thought: Einstein's and Bohr's. Not surprisingly, Einstein believed that the secrets of quantum physics would emerge from an exploration of the wave-particle duality, following the trail of his theory of light quanta. Bohr, also not surprisingly, would instead focus on the electron's discontinuous jumps between atomic orbits.

In 1924, Louis de Broglie, a historian-turned-physicist, showed quite spectacularly that the electron's steplike orbits in Bohr's atomic model are easily understood if the electron is pictured as consisting of standing waves around the nucleus, like the ones we see when we shake a rope that is attached at the other end. In the case of the rope, the standing wave pattern appears as a result of the constructive and destructive interference between waves going and coming back along the rope. For the electron, the standing waves appear for the same reason, but now the electron wave closes on itself like an Ouroboros, the mythical serpent that swallows its own tail. Just as when we shake the rope more vigorously, the pattern of standing waves displays more peaks, an electron at higher orbits corresponds to a standing wave with more peaks.

With Einstein's enthusiastic support, de Broglie boldly extended the notion of wave-particle duality from light to every moving object. Not only light but matter of *any* kind was associated with waves. In fact, de Broglie offered a formula to compute the wavelength of any chunk of matter of mass m moving with velocity v, known as "de Broglie wavelength."[4]

A baseball moving at 70 kilometers per hour (43.5 miles per hour) has an associated de Broglie wavelength of about 22 billionths of a trillionth of a trillionth of a centimeter (or 2.2×10^{-32} cm). Clearly, not much is waving there, and we are justified in picturing the baseball as a solid object. In contrast, an electron moving at one-tenth of the speed of light has a wavelength about half the size of a hydrogen atom. (More precisely, half the size of the most probable distance between an electron at the ground state and the atomic nucleus.) While the wave nature of a moving baseball is irrelevant to understanding its behavior, the wave nature of the electron is essential to understanding its behavior in atoms.

Bohr, for his part, believed that efforts to picture the electron or any other quantum object as a particle or a wave were less helpful than focusing on quantities that experiments could measure, such as the energy of the atomic orbits and the frequency and intensity of the radiation atoms emitted. In 1925, Werner Heisenberg and, a little later, Max Born and Pascual Jordan offered a formulation of the behavior of atoms strictly in line with Bohr's philosophy.

Their theory, known as "matrix mechanics," pushed aside classical notions such as the deterministic motion of particles and waves to focus on the energies between the orbits and the properties of the radiation electrons absorbed and emitted during the transitions. It was a theory of an alien world, where entities without a firm physical picture oscillated between possible states with certain probabilities. The energy difference between the orbits directly determined the frequency of the oscillations. To obtain his result, Heisenberg constructed a picture of the electron as an entity smeared in space and thus without proper location or velocity (momentum). The calculations were hard going, but the results matched the experiments. The bizarre nature of the quantum world forced physicists to invent a whole new way to describe physical reality. Deep in the heart of matter were things that were not matter, at least not in the common sense of bulky stuff. Atomism had come a very long way since Leucippus and Democritus, or Boyle and Newton. To understand Nature in its essence, Nature had to be reinvented.

For these reasons, Erwin Schrödinger's competing version of quantum mechanics, published early in 1926, was quickly celebrated as a major achievement. Contrary to the abstract matrix mechanics of Heisenberg, Born, and Jordan, Schrödinger's formulation was based on a wave equation, a much more familiar and treatable approach, aligned with the Einstein–de Broglie philosophy of associating the wave-particle duality and not discontinuous jumps as the essence of the quantum world. Initially, there was hope that quantum mechanics was deterministic after all, that it could be formulated in such a way that the future follows strictly from the past, as was the case for Newtonian mechanics, without any room for probabilities: if we know the position and velocity of a particle in a given moment of time, and the forces that act upon it, we can determine its future behavior with certainty. Excitement grew even more when Schrödinger, in the fourth and last paper of an amazingly creative series, proved the equivalence between his and Heisenberg's approaches, showing that they were two different ways of describing the same things. Schrödinger's wave equation became the entry to quantum mechanics and the physics of atoms and molecules. It is the essence of every quantum mechanics course offered around the globe.

Behind the controversy and the enthusiastic reception of Schrödinger's wave equation was the hope that Bohr and Heisenberg were wrong, that the bizarre nature of quantum physics was not fundamental but simply an expression of our incomplete understanding of Nature. Einstein, Planck, Schrödinger, and de Broglie believed that an ordered reality, perfectly deterministic, existed underlying the strange quantum world of probabilities and uncertainties. This is why, in a letter to Max Born dated December 4, 1926, Einstein famously wrote: "Quantum mechanics is certainly imposing. But an inner voice tells me that it is not yet the real thing. The theory says a lot, but does not really bring us any closer to the secret of the Old One. I, at any rate, am convinced that *He* is not playing at dice."[5] And that is also why Bohr, at the Fifth Solvay Conference in October 1927, advised Einstein to "stop telling God what to do."

Indeed, Einstein's hope and that of the so-called scientific realists was to remain unrealized. In 1927, Heisenberg showed that uncertainty was at the very heart of quantum physics, in particular in relation to position and momentum (or velocity, at least for motions with speeds much smaller than the speed of light): even with the best of instruments, an experiment cannot determine both the position and the velocity of a particle with arbitrarily high precision. In other words, we cannot know exactly where a particle is and what velocity it has, the two preconditions for a deterministic prediction of its future behavior. In light of the wave-particle duality, this is not unexpected. If an entity is neither wave nor particle but something in between (or something else entirely), it is hard to know where it is and how fast it's moving. And the smaller the object, the worse things become, consistent with the de Broglie wavelength (see note 4): we can affirm with confidence that the baseball has a certain position and speed (within the accuracy of the measurement), but not so for electrons and other small things.

Perhaps the most puzzling aspect of Heisenberg's principle is that the uncertainty inherent in quantum physics is not a technological problem resulting from instruments with limited precision; quantum uncertainty is fundamentally an expression of how Nature behaves at the shortest distances, an expression of a world alien to our own. We can't make it go away with better technology. Quite the contrary, since to measure is to interfere, the harder we try, the more we influence what we are trying to measure, and the more it squirms away from us. Like a classroom full of first graders, the quantum realm is all jitteriness. Try as we might, we can't make it stand still. As Austrian physicist Anton Zeilinger wrote in *Dance of the Photons*: "We have tried for centuries to look deeper and deeper into finding causes and explanations, and suddenly, when we go to the very depths, to the behavior of individual particles of individual quanta, we find that this search for a cause comes to an end. There is no cause. In my eyes, this fundamental indeterminateness of the universe has not really been integrated into our worldview yet."[6]

CHAPTER 22

THE TALE OF THE
INTREPID ANTHROPOLOGIST

*(Wherein an allegory explores the role of the observer in
quantum physics and how measurements interfere
with what is measured)*

Here is a little fictional tale that can help us grasp how the act
of observing influences what is being observed. There was
once an intrepid cultural anthropologist who spent years searching
for a lost Amazon tribe, mentioned in passing in a centuries-old
letter from an obscure Portuguese explorer. The letter was vague
about the tribe's location, and the explorer had disappeared without a trace. Ridiculed by his colleagues, the anthropologist—let's
call him Werner—kept on with his exploits, convinced that in such
a vast forest there ought to be unknown peoples—if not the ones
mentioned in the letter, perhaps others. "You only find if you look,"
he told his doubting colleagues back home.

After many false starts, wrong turns, and endless months of
trekking in the farthest northwestern corners of the Amazon forest, Werner saw a small clearing among a secluded grove of Carapa
trees. Werner discerned a village of about twenty huts, almost invisible to the eye, so integrated it was to its environs. A handful of
naked children ran about, kicking some kind of round seedpod.
Werner smiled: "Even here they play soccer." Knowing that the

175

natives would spot him in a matter of minutes, Werner studied his surroundings trying to find a hideout. He climbed a nearby tree and set up his sleeping bag on a flat branch, making sure there were no anacondas or other unpleasant neighbors around. The waves of flies were quite enough. He confirmed that he had water and food for about three days.

Werner picked up his binoculars and notebook and started his observations. As with other tribes, the women stayed mostly in the village, occupied with basket weaving, planting, and child rearing. Boys and men made weapons and went hunting and fishing. The whole village functioned as a unit, everyone doing a bit of everything. Motion was incessant. An elder and his wife sat under the awning of the largest hut, quietly observing the activities. "This could well be a single family, a clan," Werner thought. He marveled at the fact that he had been the first white man to actually observe them in their pristine state. "Everything they need they have right here; the forest gives them it all. It's actually hard to draw a line between humans and forest; they all behave in perfect integration."

A boy, probably five or so, fell down and scraped his knee rather badly. The elder woman rushed to him and rubbed some ointment on the wound. The boy smiled and went back to play, apparently without any pain. "The elder woman is clearly the tribe's physician," Werner wrote in his book. "I need to find out what plant she used to anesthetize the boy's pain."

At night, after the men and the older boys had returned from their foray, the whole tribe met around a central fire. The elder man told some sort of story, probably a tale of days long gone. At the end of every sentence, the tribe would sing together some kind of mantra in celebration of the deeds of their ancestors.

Werner went to sleep only after they were all inside their huts. "What phenomenal luck," he whispered to himself. "The idiots back home are going to eat their hats!" Werner closed his eyes the happiest man on Earth. He was just about to doze off when someone shook him by the shoulders. He had been found out! Three large men took him down from the tree and brought him into the elder's tent. There was a lot of shouting and pointing. They stripped him naked and carefully examined his clothes and body. They were studying him as he had studied them. "If I survive this, I promise to do everything I can to protect this village," he thought. To his surprise, the elder woman brought in some kind of hot beverage, gesturing for him to drink it. Werner did as he was told. In a few minutes, he was asleep.

The Sun was high in the sky when he woke up. The natives had set up a hut next to that of the elders, expecting him to join them. Werner was delighted. "So I'm not dead yet. This will allow me to continue my observations," he thought. And so he did. It didn't take him long to realize that the whole clan dynamics had changed. He had become the focal point for young and old alike. Children wouldn't leave him alone, pulling on his grey beard and asking him to kick the seedpod with them. The younger women eyed him with lust, wondering how different it would be to have sex with a man of pale skin. The warriors were always on guard, expecting Werner to attack them at any moment. "They are not the same, and never will be," Werner realized with sadness. "My presence here has altered their behavior in irreversible ways. I destroyed their worldview and forced them into a new one that can't be changed." Werner had changed too. He wasn't sure he wanted to go back home.

I tell Werner's story to illustrate the difference between a classical and a quantum approach to measurement. Before being found out, Werner's information about the clan was "pristine," that is, unspoiled by his presence. This is the ideal situation for an observer, where the act of observing doesn't affect what is being observed. A detachment, a separation between the two, is preserved. Our perception of reality is very much based on this kind of measurement, given that we are conscious only of large objects for which quantum effects are, it seems, of small importance. We see books piled up on the desk, cars passing by on the street, flies buzzing around, and our observing of these objects does not affect their behavior. (Of course, if you make a move toward the cars or the flies, they will respond. But this is not the point here.) This is the classical approximation, the limit of quantum physics at which quantum effects don't play a role, the world of our sensorial perception. As we shall see, carefully examining whether such classical approximation is realistic—even if we may think so, given that it is based on what we perceive of the world—will teach us a lot about the nature of quantum physics.

The other situation, the tribe after Werner's presence was discovered, illustrates the quantum realm, where the act of observing interferes with and irreversibly changes what is being observed and the observer. The natives were never the same after they had found Werner. Neither was he. After he was discovered, Werner became part of the tribe and the tribe part of Werner: they formed an indissoluble whole. Their experience of each other's existence affected their histories in irreversible ways; neither Werner nor the tribe would ever revert back to their prediscovery, mutually independent states. We say that they became "entangled" as a result of their mutual interaction, a term Schrödinger used in this context for the first time in a paper from 1935, claiming it was *the* essential property of quantum systems.

CHAPTER 23

WHAT WAVES IN THE QUANTUM REALM?

(Wherein we explore Max Born's bizarre interpretation
of quantum mechanics and how it complicates
our notion of physical reality)

Let's regroup for a moment. The first twenty-five years of the twentieth century were the stage of a triumphal success in our description of the physical world: Einstein and his theories of relativity, and the quantum mechanics of Heisenberg, Schrödinger, and others.

The unusual aspect of quantum mechanics is not related to its efficiency as a physical theory: it is the most successful theory we have of Nature, able to describe with amazing accuracy the properties of countless materials, molecules, atoms, and subatomic particles. The challenge with quantum physics is with its interpretation, of making sense of what's "really" going on. We have already seen how small objects show both wave and particle behavior depending on the experimental apparatus that they are subjected to. We saw that this dual behavior results from an intrinsic indeterminacy in Nature, summed up in Heisenberg's Uncertainty Principle. We saw that a consequence of this principle is that observer and observed cannot be separated, since the very act of observing affects what is being observed—in fact, more than affects, and this is a contentious point, it *determines* what is being observed. Let me say it differently: if quantum mechanics is right, and there is no indication that it isn't, the observer determines the physical nature of what is being observed. An electron is neither wave nor particle; it *becomes* one or the other depending on how it is observed. In a collision experiment an electron behaves like a particle; make it go through two narrow slits, and it creates an interference pattern like a wave. In the quantum world everything is potentiality, a lottery in which outcomes depend on who (or what) is running the show.

"But surely," scientific realists would object, "things in Nature must be something before we observe them; they *must* exist in some form."

"Maybe," a strict quantum mechanic would reply, "but we can't tell what that something is, and it really doesn't matter; what matters is that we can explain our experiments with this construction, even if it is bizarre."

"Are you telling me that things only exist when we look at them? That the electron is really not there until we interact with it?"

"Yes, that's exactly what I am saying. For all practical purposes, the electron only exists when we measure it."

"But what of bigger things? After all, rocks are made of atoms, trees are made of atoms, people are made of atoms. Are these also not there until we look at them?"

"Strictly speaking, that's right. We don't know if something exists until we interact with it. Even big things: we *assume* that they are there because they were there before, but we can't be sure until we look. But in practice most people imagine that there must be a dividing line or, better, a transition region where classicality takes over as an effective description of physical reality. This is explained using something called 'decoherence,' which we will explore a bit later."

"Of course. But no one knows how to define this transition region, right? In theory at least, nothing exists until observed."

"Sounds silly, I know. That's why we don't like to go there. We use quantum mechanics as needed, compute with it, and go home."

"That's okay if you don't care about truly understanding the nature of reality, if you are content with just computing stuff. Don't you want to go beyond the 'for all practical purposes' prison?'"

"Well, perhaps the lesson from quantum mechanics is simply that we can't understand the nature of reality, that we must learn to live with this realization and accept that we can only have partial knowledge of what reality is. We must learn to let go."

"I don't buy it. What about the Universe itself? Wasn't it small near the Big Bang? Wasn't it a quantum object then? If it was, and if everything is quantum mechanical, shouldn't the Universe still be a quantum object? Or is it classical now? Who is observing it?"

"Sorry, friend, this is when I go home."

"*Esse est percipi.* Isn't this what Bishop George Berkeley said in 1710, 'To be is to be perceived'?"

"Yes, but Berkeley meant this as a proof for the existence of God, the eternal observer who thus gives reality to all things. I don't think this helps us understand quantum mechanics."

"The mysteries of existence and of the nature of reality get entangled . . . "

"Okay, I'm *really* going home now!"

The elusive quantum behavior is not particular to the very small. Everything is quantum mechanical; everything is jittery and uncertain. The difference is that for small objects the jitteriness has a huge impact, while large objects have such small quantum jitters as to appear to have none at all. The point is that the world is quantum mechanical, not classical. The Newtonian worldview is a very successful approximation to what goes on for large objects, when it is "safe" to neglect any quantum effects. Still, it's only an approximation. Although there are ways to estimate when quantum effects become unimportant (for example, a small de Broglie wavelength compared to the dimensions of the system, high temperatures, strong environmental influence), some quantum effects persist to surprisingly large scales. Is the Universe as a whole quantum mechanical?[1]

When Schrödinger wrote his wave equation, he had to figure out what it meant, give it an interpretation. Something was waving in space and time, something called the "wavefunction," represented in his equation by a mathematical function of space and time, $\psi(t,x)$. Since his first application was to explain de Broglie's electron waves in the different orbits of the hydrogen atom, Schrödinger's natural guess was that these were "electron waves." As this first attempt didn't quite work, he then tried to interpret his wave equation as describing the charge density of the electron: picturing the electron as some sort of diffuse wave of charge, the solution to the wave equation would give the electron's most probable location. In a letter to Hendrik Lorentz dated June 6, 1926, Schrödinger got tantalizingly close to the right interpretation, suggesting that maybe it was the square of the wavefunction that should be interpreted physically: "the physical meaning belongs not to the quantity itself but rather to a *quadratic* function of it."[2] For the cognoscenti, the quantity relates to the absolute square of $\psi(t,x)$, since it is a complex function. (More on this in note 3.)

Schrödinger had no intention of abandoning the notion that his equation described something concrete. He was really close to getting the correct interpretation but couldn't let go of his scientific realism. A few days after Schrödinger published his paper attempting to interpret the wavefunction as the electron's density of charge, Max Born came up with an alternative which, to Schrödinger, de Broglie, Planck, and Einstein, was nothing short of outrageous: the wavefunction did not represent an electron or its charge density. In fact, it was nothing real. The mathematical function figuring in Schrödinger's equation, suggested Born, was a computational device. Its role was to give information about finding the electron in a given place at a given time with a certain energy, called a "probability amplitude." By squaring it properly one obtains the "probability density," a number between zero and one that gives the probability that a measurement of the electron's position will find it in the neighborhood of a certain position x.[3]

It's important to understand that the wavefunction gives the probability amplitude of the electron being somewhere *before* the measurement takes place. Once a device capable of detecting an electron is turned on and the electron is detected at some point x, it will stay there: the interference between detector and electron seals the electron's fate. Schrödinger's equation gives the probability that the electron will be found somewhere, but once it is, the drama is over. The wavefunction "collapses" entirely into one point in space (within the accuracy of the measurement). In ways that remain obscure, detection selects a specific spot for the electron to be. The wavefunction gives the probability for it being found here or there, but the actual spot where it is found is unknown. Even more strangely, the collapse happens instantaneously: whereas the wavefunction was spread out in space before the measurement, it immediately collapses to the neighborhood of a point with the measurement. This goes against the notion of "locality," the requirement that no physical influence can travel faster than light: only causes that had enough time to reach an object (and are thus local in this sense) can influence its behavior. How does the wavefunction, its different "parts" at faraway distances, "know" to collapse as a whole and instantaneously?[4]

Maybe an analogy would help. Niels was doing some construction work in Werner's home in the outskirts of Manaus, the capital of the state of Amazonas in Brazil. (Werner didn't go back to Germany after all, but he didn't stay in the jungle either.) Werner lived in a very secluded place, dense forest all around. Niels left for a few minutes, leaving the window open. Within minutes, a large surucucu, the deadliest in the long list of deadly Brazilian poisonous snakes, sneaked in and decided to nap stretched over many steps of a wide ladder. Upon noticing the open window, Werner anticipated trouble and tiptoed outside the house. He peeked in and saw the happily resting ten-foot-long monster. Hardly breathing, Werner started to look for a long stick when Niels came back. "Hey Werner, what's going on?" "Shhh! Do you want to die? Look!" When the snake sensed the noise, it tensed its body and in a flash wound itself around one rung of the ladder. From being in no specific rung while napping, the snake "collapsed" into one after interference from Niels's voice. And it didn't seem very pleased to have its nap interrupted.[5]

The mathematical apparatus of quantum mechanics describes what matter does without referring directly to it. The equation mentions the forces that act on the electron (or whatever particle or atom under study), but not the electron itself. It mixes "real" things (forces and energy) with nonreal things (the wavefunction). The wavefunction contains all the statistical information we can get from a physical system, but it does not directly represent the entities that are part of the system. In contrast, when someone throws a rock into a pond, we can write an equation describing how water waves propagate. The solution to this equation is a mathematical function that represents a real water wave: there is a one-to-one correspondence between the water wave and the mathematical function that describes how it propagates. In contrast, the "wavefunction" in Schrödinger's equation is a mathematical function that doesn't describe how something real is waving.

This unusual conceptual structure invites the question as to what, exactly, are the "entities" being described and where were they before the measurement was made. In our analogy, the surucucu was "everywhere," spread across the ladder. But a snake, even if in principle a quantum entity, is very well approximated by a classical model. We can see it before and after it notices us. It also doesn't "collapse" to one rung instantaneously but does it following well-behaved causal steps: sensorial perception of a potential pray triggers nerve impulses that induce muscular contractions into attack mode. Are entities in the quantum realm real in the same sense that the snake is real? Could we and rocks and lakes be made of entities that are not real? The question of what is real is at the very core of how we interpret quantum mechanics.

Physicists have been wrestling with this question since the early days of quantum mechanics. Einstein, Schrödinger, and the scientific realists supposed that the current description was a temporary placeholder, waiting for a more complete explanation to develop. Notice the choice of the word "description" versus "explanation." Scientific realism supposes that science *explains* what's real, that real entities exist at all levels, and that explanations, when available, reach them all. What mostly bothered Einstein about quantum physics was not so much its probabilistic nature but its disdain for the whole reality program of science. "Is the Moon not there if I'm not looking?" he half-jokingly asked a friend during a walk in Princeton.

Einstein's quip apart (we will get back to it soon), *explaining* reality may be too much of a tall order, even for science. Especially if we attach to an explanation some sort of finality, which, as I have argued, is incompatible with the way science advances. The aether, the phlogiston, the caloric, and even Bohr's model of the atom all functioned as descriptions of natural phenomena, rightly or wrongly. They performed a useful function while they were considered, working as bridges between old and new ideas. Clearly, there was no physical reality to any of them in the sense that scientific realism would have liked. More appropriate to the scientific enterprise with its ever-shifting perspectives is to consider our models and theories as *descriptions* of the portions of reality that we are able to measure and make sense of. When it comes to physical reality, there are no final explanations but ever more efficient descriptions.

An example of the realist philosophy is the de Broglie−Bohm hidden-variable theory (developed in detail when David Bohm worked with Einstein in Princeton and then when he went to São Paulo in 1952 to escape McCarthyism), which resonates with early ideas by de Broglie. Bohm added an extra level of explanation to quantum theory capable of describing the electron's position with certainty, called the "pilot wavefunction." While Schrödinger's equation remained the same, another equation would work as its "pilot": just as a conductor controls how different sections of an orchestra play during a symphony, the pilot wavefunction determined how the wavefunction branched out into different probable states. This conducting happened through one or more undetectable hidden variables, information that remained out of reach of experiments. The pilot wave acted everywhere at once like an omnipresent deity, a property physicists call "nonlocality." In other words, in the de Broglie—Bohm mechanics, particles remained particles, and their collective motion was guided deterministically through the nonlocal action of the pilot wave. The particles were like a group of surfers gliding along a single wave, each pushed this way or that as the omnipresent wave advanced.

In the de Broglie—Bohm theory the behavior of the electron was perfectly predictable, such that finding it here or there was calculable with accuracy. The hidden variable would be the missing link between a classical concept of reality and the fuzzy world of quantum indeterminacy. The price of making quantum mechanics deterministic was to impose an endless web of influence between everything that exists: in principle, the whole Universe participates in determining the outcome of every experiment. In practice, the velocity and acceleration of each particle depend on the instantaneous positions of all other particles. The Universe acts in tandem to determine the environmental conditions that apply to each and every subsystem, from a collider experiment to the motion of clouds in the sky. This is what physicists call "nonlocality," but with a vengeance. No wonder Bohm's book exploring the philosophical foundations of his ideas is called *Wholeness and the Implicate Order*. And no wonder few physicists endorsed his approach, although some variants of the de Broglie—Bohm theory are still the subject of active research. One of the issues is that the theory (at least most versions of it) gives the same results as quantum mechanics, and hence is indistinguishable from it: the hidden variables are undetectable. We would like to be able to distinguish between competing theories on the basis of experiments; if different theories give the same experimental results, why not pick the simplest one, that is, traditional quantum mechanics without extra pilot waves? I will thus leave hidden variables aside for now and concentrate on what quantum mechanics is or isn't telling us about the nature of physical reality.

CHAPTER 24

CAN WE KNOW WHAT IS REAL?

*(Wherein we explore the implications of quantum physics
for our understanding of reality)*

One of the shocking consequences of quantum physics is that the act of measuring affects what is being measured. In fact, it defines what is being measured, giving it physical reality. This creates a link between observer and observed that is hard to sever. Perhaps no one put it more pointedly than Pascual Jordan, who worked with Heisenberg and Born in the formulation of matrix mechanics: "Observations not only disturb what has to be measured, they produce it. . . . We compel [the electron] to assume a definite position. . . . We ourselves produce the results of measurements."[1]

Once the link exists, the separation between you as observer and the rest of the world, what we usually call objectivity, is lost. How can you know where you end and what you measure begins? If we are entangled with what is "out there," there is no "out there" any more; there is only the whole of it, undifferentiated. Detachment is gone. You and everything else in the Universe make up a single unit. Even more problematic: If you are connected to everything else, to what extent are you free? Is our autonomy as individuals an illusion? Does the sum total of the influences out there dictate our behavior? Are we the spider that can't exist without the web?

"Surely," someone with a cool head may object, "that's not how things are in real life. Just look around, and you can see that we are apart from what's out there, that we exist independently of it. I'm not the chair I'm sitting on. The chair has its own existence, independent of my own. It is an autonomous object with nothing quantum about it. Furthermore, you don't detect the particle; a machine does, the detector. And the detector is also a large, classical object. So the extrapolation that the act of measurement connects observer and observed is a bit of a stretch. What happens is that a particle interacts with the materials that make up the detector, and this interaction, after being sufficiently amplified, is recorded electronically in a counting or tracking device. There is no 'you' or consciousness or mind behind the particle's existence; there are just clicks in detectors. The business of quantum mechanics is to make sense of these clicks, and it does so beautifully using probability. Microscopic objects don't exist in the same way you and I exist; they are just constructions of our minds, descriptive devices we create to make sense of what we measure. Why go all metaphysical with it?"

The above paragraph reflects what sometimes is called the "orthodox" position, based on the Copenhagen Interpretation of quantum mechanics that Bohr and Heisenberg originally developed to assuage confusion and despair. When we teach quantum mechanics, we tend to remain within the confines of the Copenhagen Interpretation and its pragmatic approach to reality. This is an acceptable position as long as you don't want to go deeper into the nature of things. But as soon as we start to think a bit more about it, an unsettling feeling creeps in. And the feeling only grows as our thoughts deepen.

It is certainly true that it is a detector that signals the existence of the particle, not a person directly. But the scientist and his intentionality, that is, his specific design for the experimental setup, comes before the detector. A detector doesn't exist without the scientist and won't work without someone turning it on or programming it to turn on at a certain time. The data the detector collects won't make sense without a conscious observer who knows the science behind it. An electron doesn't really exist without a conscious mind to interpret it. Put it another way: existence, be it of a quantum or of a "classical" object, is contingent on minds to acknowledge it. In a mindless Universe nothing exists, since there are no conscious entities aware of what existence even means. The very concept of existence presupposes a mind capable of higher reasoning: "existence" as a concept is something we invented to make sense of how we fit in the cosmos.

Of course, this doesn't mean that the Universe only came into existence once there were conscious observers to notice it. Unless you agree with Bishop Berkeley and his *Esse est percipi*, the Universe existed long before any conscious observers came about. Humans and any other intelligence out there capable of thinking about existence are the result of countless physical and chemical interactions that engendered, in ways that remain unclear, complex biological entities. This all takes time, no less than a few billion years, enough for several generations of stars to come into being and perish, cooking up the heavier chemical elements of the periodic table that are crucial for life. Given that there were no minds at the beginning of time, we must conclude that consciousness is not a precondition for the Universe to be.[2] Indeed, if the multiverse makes sense, countless universes may exist out there without any trace of life in them. The opposite is obviously not true: life unfolds within a universe. Unless you believe in some sort of universal disembodied Mind, life presupposes a complex web of physical, chemical, and astronomical conditions operating within space and time. Many eons of cosmic history passed before life could start having a history of its own.

The key question then is not whether consciousness engenders the Universe—a very difficult position to defend scientifically—but rather what happens to the Universe once consciousness emerges. You may dismiss this with a Copernican flourish, arguing that we are negligible in the big scheme of things, that we came from stardust and to stardust will revert. In response, I'd argue that the Copernican position hinges on the wrong axis: what matters isn't whether the Universe cares about us, for it clearly doesn't. What matters is how *we* fit into the Universe once we understand our uniqueness as conscious beings. I called this position "humancentrism" in *A Tear at the Edge of Creation*. In a nutshell, we matter because we are rare. Even if there are other "minds" in the cosmos, we are a one-of-a-kind experiment in evolution.

How does this relate to the foundations of quantum physics and the nature of reality? For starters, everything that we can say about reality goes through our brain. When we design the experiment that determines whether the electron is a particle or a wave, "we" means the human brain and its ability to reason. Detectors are extensions of our senses designed to record events that we then decode through careful rational analysis. We have no direct contact with electrons, atoms, or other denizens of the realm where quantum effects prevail; all we get are flashes and pings and ticks and lines and reams of data that we rush to interpret. The world of the very small exposes in direct ways the limitations of our descriptions of reality. Yet these descriptions are all we've got. As such, they reflect in very deep ways our human essence, the ways we pursue knowledge and our limitations in doing so. We are meaning-seeking beings, and science is one offspring of our perennial urge to make sense of existence.

Even though I have used quantum mechanics in my research for decades, and have taught quantum mechanics and quantum field theory for as many years, as I began to survey the literature on the conflicting interpretations of quantum mechanics, a sense of loss took over my thoughts like a hungry vine. Could reality be this elusive? The hardest part is that there is no simple resolution, no agreed-on way out. Even though we all go through the same motions when calculating quantum probabilities, there is widespread disagreement as to how quantum mechanics relates to reality. There may not *be* a correct resolution—only different ways to think about it. The difficulty, as we will see next, is that some quantum effects force us to confront their weirdness in ways that affect how we relate to the Universe. Could it be that there is no "us and the Universe" but a single wholeness? Unless you are intellectually numb, you can't escape the allure of the quantum, the tantalizing possibility that we are immersed in mystery, forever bound within the shores of the Island of Knowledge. Unless you are intellectually numb, you can't escape the awe-inspiring feeling that the essence of reality is unknowable.

In 1935, Einstein published a paper with Boris Podolsky and Nathan Rosen (referred to below as EPR) trying to expose the absurdities of quantum mechanics. The title says it all: "Can [the] Quantum-Mechanical Description of Physical Reality Be Considered Complete?"[3] The authors had no qualms with the correctness of the quantum theory: "The correctness of the theory is judged by the degree of agreement between the conclusions of the theory and human experience. This experience, which alone enables us to make inferences about reality, in physics takes the form of experiment and measurement." Their issue was with the completeness of the quantum description of the world. They thus proposed an operational criterion to determine the elements of our perceived physical reality: those physical quantities that could be predicted with certainty (a probability of one) without disturbing the system. That is, there should be a physical reality that is entirely independent of how we probe it. For example, your height and weight are elements of physical reality as they can be measured with certainty (within the precision of the measuring device). They also can be measured simultaneously, at least in principle, without any mutual interference: you don't gain or lose weight when your height is measured. When quantum effects dominate, this clean separation is not possible for certain very important pairs of quantities, as expressed in Heisenberg's Uncertainty Principle. EPR would have none of this.

We have seen that the uncertainty relation precludes knowledge of both position and velocity (momentum, really) of a particle. This is true for many different pairs of quantities said to be "incompatible." Energy and time are also incompatible, and they obey an uncertainty relation similar to that for position and momentum. Another example is the spin of a particle, a quantum property that we associate with some kind of intrinsic rotation and visualize, even if incorrectly, as the particle turning around like a top. Quantum particles with spin are like whirling dervishes who can never stop. Not only that, they always whirl with the same speed (angular velocity), although different particles may have different spins. Spins in different directions (say, aligned in the north-south or east-west directions) are incompatible: we can't measure them simultaneously. Classically this limitation usually doesn't exist, since most physical quantities are compatible.[4]

When quantities are compatible, you can obtain information about both without any a priori restriction. In quantum physics, whenever two quantities are incompatible, the uncertainty principle applies: the information we can obtain about both of them jointly is restricted. If we know the momentum of a particle and also want to know the position, a measurement of the position will "force" the particle into a specific spot, "collapsing" its wavefunction: in other words, the measurement will decisively disturb the particle and change its original state. More dramatically, we can't even speak of an "original position state": before the measurement, all that existed were potentialities of the particle being here or there.

Back to the EPR paper. We see that incompatible quantities violate their proposed criterion for a physical variable to belong to physical reality: since measuring the particle's property means disturbing it, the act of measurement compromises the notion of an observer-independent reality. The act of measurement *creates* the reality of a particle being in a given spot in space, which they found absurd. What is real must not depend on who or what is looking.

EPR considered a pair of identical particles moving with the same speed in opposite directions. Call them particles A and B. Their physical properties were fixed when they interacted for a certain time before flying off away from each other.[5] Say a detector measures the position of particle A. Since the particles have the same speeds, we also know where particle B is. But if a detector measures particle B's speed at that spot, we now know *both* its position and its speed. This seemed to clash with Heisenberg's Uncertainty Principle, since information was apparently obtained about a particle's position and velocity simultaneously. Furthermore, we know a property of a particle (position of B) without observing it. According to the EPR definition, this property is then part of physical reality even if quantum physics insists that we could not know it before we measure it. Clearly, argued EPR, that doesn't seem to be the case, and quantum mechanics must be an incomplete theory of physical reality. EPR closed the article hoping that a better (more complete) theory would restore realism to physics.

Bohr's answer came after only six weeks, in a paper he provocatively titled the same as EPRs. (I don't think you could do this nowadays.) Bohr invoked his notion of "complementarity," which asserts that in the quantum world we cannot separate what is detected from the detector: the interaction of the particle with the detector induces an uncertainty in the particle but also in the detector, since the two are correlated in inseparable ways. Essentially, the act of measurement *establishes* the measured property of the particle in unpredictable ways. Before the measurement we can't say it had any property at all. This being the case, we also can't attribute physical reality to this property in the sense that EPR defined: "Indeed the *finite interaction between object and measuring agencies* . . . entails . . . the necessity of final renunciation of the classical ideal of causality and a radical revision of our attitude towards the problem of physical reality"[6] (italics in the original).

In his classic textbook, David Bohm elaborates: "[We assume that] the properties of a given system exist, in general, only in an imprecisely defined form, and that on a more accurate level they are not really well-defined properties at all, but instead only potentialities, which are more definitely realized in interaction with an appropriate classical system, such as a measuring apparatus."[7] Bohm then takes it home, dialing up the dramatic rhetoric: "We see then that the properties of position and momentum are not only incompletely defined and opposing potentialities, but also that in a very accurate description they cannot be regarded as belonging to the electron alone; for the realization of these potentialities depends as much on the systems with which it interacts as on the electron itself."[8]

According to Bohr and his followers, EPR implicitly built their arguments using the old classical assumption that there is such a thing as a reality independent of measurement. That expectation, they insisted, had to go. Reality was far stranger than Einstein would have liked it to be. All we could do was to probe it with our measuring devices the best way possible and make sense of the results using the probabilistic interpretation of quantum mechanics. What lies underneath, if anything, was *unknowable*. This is why Heisenberg wrote, "What we observe is not Nature itself but Nature exposed to our method of questioning."

In EPR we identify traces of Plato's idealism, the notion that there is an ultimate reality out there, the underlying stratum of all there is, and that it is accessible to reason. The key difference is that while for Plato this reality was in the abstract realm of Ideal Forms, for Einstein and the scientific realists it was very concrete, even if hard to get to. The clashes with the pragmatism of the Copenhagen Interpretation and with Bohr's complementarity was direct and unavoidable.

Were Einstein, Schrödinger, and the scientific realists being merely hopeful, echoing ancient dreams of a complete understanding of the world? How far *can* we go unveiling Nature's underlying structure, as opposed to seeing only shadows on the wall? Is the underlying stratum of physical reality truly unknowable?

Schrödinger would have none of this. In 1935, prompted by the EPR paper and Bohr's response, he composed his own critique of quantum physics in which he introduced his famous cat. Schrödinger's intent was to ridicule the very theory he helped to found when it is extrapolated to macroscopic objects. He had a point.

Consider a cat locked in a black box together with, as Schrödinger called it, a "hellish contraption": a Geiger counter attached to a sample of radioactive atoms and a bottle of cyanide. When an atom decays and emits a particle, the Geiger counter detects it, triggering a mechanism that releases cyanide from the bottle, killing the cat; if the atom doesn't decay, the cat stays alive. Clearly, an outside observer can't tell whether the cat is alive or dead until he opens the box. Schrödinger's point is that quantum mechanics would state that *before* the box is opened, the cat is *both* alive and dead. The wavefunction describing the whole system would contain equal parts of a living and a dead cat. (It would be in a "superposition" of both states.)[9]

According to the Copenhagen Interpretation, the act of looking has a 50 percent probability of killing the cat. Talk about looks that can kill! And that's not all: if the cat is either dead or alive when you open the box, its past history must reflect this—that is, it was or was not poisoned. Does this mean that the act of observing actually determines the past history, acting backwards in time? Can a look not just kill but recreate the past?

One response is that the measuring device is the Geiger counter and not the person opening the box: if the atom decays and the Geiger counter registers the decay, this constitutes the act of measurement. You may counter by arguing that since we don't know what goes on inside the box, what happens between the cat and the Geiger counter is irrelevant. Only looking has meaning, since it explicitly introduces the observer into the picture.

At the heart of the puzzle lies a paradox that was inexistent in classical physics. In quantum physics, the trio consisting of the observer, the measuring device, and what is being measured form a new entity, which is described by a single wavefunction. As Schrödinger explained, their individual wavefunctions are "entangled."[10] In principle, the whole Universe should be part of the description, given that all sorts of remote effects act on all of us: Jupiter's gravity, the Sun's radiation, the pull from the giant black hole at the center of the Milky Way and the one at the center of the Andromeda galaxy, the bird fluttering its wings outside the window and the clouds drifting across the sky, the waves crashing at Ipanema beach, and so forth. How can we reconcile this entangled universal wholeness with the fact that the act of observation necessitates that what is being observed is *distinct* from who (or what) is observing? Otherwise, if observer and observed can't be separated, how do we know where one ends and the other begins? Isn't this separation the essence of measurement?

Fortunately, the vast majority of measurements are such that the small quantum effects coming from the interactions between the observer and his apparatus or between the observer and the rest of the Universe are perfectly negligible. Their net statistical impact is much smaller than the typical experimental errors arising from limitations of the measuring devices. We are thus justified in treating the observer and her measuring device as two distinct entities interacting strictly along the laws of classical physics. Also, since the states of the measuring device are the same for different human observers (clicks on a Geiger counter, deflections of a measuring gauge, tracks on a cloud chamber etc.), we are justified in considering these states as independent from the act of observation or of the particulars of the observer. The quantum theory of measurement conveniently reduces to analyzing data collected from a classical device designed to capture and amplify signals from an observed system. This description should be effective as long as there is a clear separation of scales so that the measuring device behaves classically.

This sharp distinction between what is observed and the measuring device, which is at the very core of Bohr's complementarity idea, made sense sixty years ago, when the difference in scales was indeed huge. However, experiments today probe the "mesoscopic" realm, the mysterious boundary between an acceptable classical description and quantum physics, roughly below one-millionth of a meter, the size of a bacterium. Atoms can be visualized and manipulated individually, as in the famous 1989 IBM experiment in which Don Eigler used a scanning tunneling microscope to manipulate thirty-five argon individually and construct the company's initials. Nanotechnology explores the fabrication of devices at the mesoscopic scale, taking advantage of quantum effects. Certain contraptions are sensitive enough to capture oscillations coming from the "zero-point energy" of quantum harmonic oscillators, effectively detecting the energy of the vacuum. Far from being a shortcoming, the elusiveness of the quantum realm is being put to work in the development of revolutionary new technologies, from highly secure bank wirings to ultrasensitive detectors and, potentially, new types of computers.

The net result is that the boundary between the quantum and the classical is no longer well defined. In many applications physicists can't hide behind Bohr's conveniently pragmatic separation between a quantum system and its classical measuring device. The weirdness must be faced head-on. This explains why so many more physicists work today on the foundations of quantum mechanics than, say, even twenty years ago.[11] The question, though, remains: Is quantum weirdness an unavoidable aspect of Nature, or can we somehow make sense of it? The answer is essential to our argument, since, if the weirdness of quantum mechanics can be explained, it would simply imply a further growth of the Island of Knowledge, while if it can't, we would have to accept that large portions of physical reality are not just unknown but unknowable to us.

Critics of the Schrödinger cat puzzle would claim that a cat is just too big to isolate from the rest of the world and be placed in a superposition of two states, dead and alive. The whole thing is impractical and thus nonsense. At first glance it may be. But where do we draw the line? Austrian physicist Anton Zeilinger and his group have performed amazing experimental feats, making increasingly large objects go through double-slit obstacles to test whether they create interference patterns like electrons and photons.[12] In 1999, they succeeded in interfering buckyballs—a large spherical molecule with sixty carbon atoms and shaped like a soccer ball. More recently, they have extended their reach to include large biomolecules and intend to test if viruses can be put in a superposition of quantum states and interfere. As the object's size increases and its associated de Broglie wavelength decreases, it becomes much harder (and more expensive) to isolate objects from external influences and place them in a superposition of two or more quantum states. If a single photon emitted from the box wall bounced off the cat, it could, if it escaped and were detected, tell us whether the cat was standing or lying down. The single photon could collapse the cat's wavefunction. Still, the day will come when quantum interference experiments will attempt to pass a bacterium through double slits. How would life respond to quantum superposition? Is life a classical state of matter?

Schrödinger was well aware of these difficulties. His challenge was not experimental but conceptual. Was there a boundary between the quantum weirdness and our supposedly more reasonable conception of reality? Surely the world doesn't seem to be made of superimposed quantum states. Taking the three seminal papers from 1935—the EPR paper, Bohr's response to it, and Schrödinger's own take on the matter—we can see why most physicists opt to ignore all of this and go on with their work, happily computing transition rates and quantum superpositions as if there were nothing to worry about. But if we take a careful look at what the EPR paper was really saying, and how current experiments actually confirm the bizarre reality that it tried to deny, including faster-than-light action-at-a-distance, how can we just dismiss the whole thing as a mere philosophical debate? Einstein and Schrödinger were convinced that Nature was trying to tell us something. Perhaps we should listen more carefully—which is what we do next.

$$\bowtie\!\!\!\bowtie$$

WHO IS AFRAID OF QUANTUM GHOSTS?

*(Wherein we revisit what so bothered Einstein about
quantum physics and what it tells us about the world)*

Before you get too comfortable accepting Bohr's refutation of the EPR challenge and adopt a pragmatic approach to quantum physics, let's revisit it with a more modern setup, one that is actually realized in experiments.

When light is polarized, its associated wave goes up and down in the same direction of the polarization, as when we ride a horse. (This is the direction of the electric field characterizing the electromagnetic wave.) Photons of polarized light share this polarization. The details of how this works are not important; what matters is that the photons have this property and that it can be measured.

Imagine that in an experiment a source of light created a pair of polarized photons traveling in opposite directions, say, east and west. Imagine that two physicists, Alice and Bob, each stood with a detector at one hundred yards from the source: Alice at the left and Bob at the right. Since photons travel at the speed of light, Alice and Bob would see photons arriving at their detectors at the same time.

[ALICE] — — — — — (SOURCE) — — — — — [BOB]

Now imagine that the detectors can identify two possible polarizations for the photons, vertical and horizontal. The light source always produces pairs of photons with the *same* polarization. Alice and Bob don't know which polarization this pair has until they measure it. Let's say Alice measures vertical; Bob will measure vertical too. If Alice measures horizontal, so will Bob. Even though there is a fifty-fifty probability for the photon to be in either state of polarization (the vertical or horizontal polarization appears randomly), Alice and Bob will *always* obtain the same outcome: the two photons leaving the source are entangled. They behave as one.[1]

Alice decides to move a bit closer to the source. She measures a photon with vertical polarization. Immediately, she knows that Bob's photon will also have vertical polarization, *before the photon even gets to Bob's detector*. But according to quantum mechanics, you can only tell the state of something by looking. And since nothing can travel faster than the speed of light, Alice apparently influenced Bob's photon instantaneously without interacting with it! (If not instantaneously, at least superluminally, faster than the speed of light.)

Amazingly, the effect does not depend on how far Alice and Bob are from one another. They could have been ten miles or light-years away, and the same thing would have happened. Within the accuracy of current detectors, everything appears to happen instantaneously. Note, however, that no information was transferred between the two photons. They didn't "talk" to each other (interact with each other) in any (known) way. They behaved as a single entity perfectly impervious to spatial separation. Einstein called this effect "spooky action-at-a-distance," the mysterious and wondrous quantum ghost. Given what he had done to Newton's ghost—explaining gravity as a local influence and not as an action-at-a-distance—we can see why he was so keen on exorcizing the quantum one too. He died convinced that such spooky effects had to go. But can they?

Such entangled pairs of particles are created and analyzed in many laboratories around the world. Measuring one of the two "influences" the other instantaneously (or at least superluminally), irrespective of how far apart the two are from each other. It's time to look at these experiments in more detail.

CHAPTER 26

FOR WHOM THE BELL TOLLS

*(Wherein we discuss Bell's theorem and how its
experimental implementation shows how
reality is stranger than fiction)*

Could there still be a way out? Maybe physicists are missing something important and obvious, the "right" understanding of what is truly going on? It wouldn't be the first time, as our historical explorations have shown. Maybe what Einstein did for Newton's action-at-a-distance could be somehow replicated for the quantum case? Maybe the action-at-a-distance is not instantaneous, just superluminal? After all, we could never determine if something acts instantaneously, since it would require measurements of time intervals with absolute precision, something no measuring device can do in practice. "Instantaneous" and its opposite, "eternal," are concepts that cannot be experimentally validated. No measurement can ever be fast enough to be instantaneous or long enough to last forever. We can't ever be sure if they are part of physical reality.

We saw that Bohm had proposed a nonlocal theory of hidden variables that was in agreement with the predictions of quantum mechanics. This approach was consistent with what he wrote in his textbook, published just one year before his 1952 papers: "Until we find some real evidence for a breakdown of the general type of quantum description now in use, it seems, therefore, almost certainly of no use to search for hidden variables. Instead, the laws of probability should be regarded as fundamentally rooted in the very structure of matter."[1] So any hidden-variable theory had to duplicate the successes of quantum mechanics. In addition, and this is why Bohm developed his theory, it had to provide a "precise, rational, and objective description of individual systems at a quantum level of accuracy."[2]

For twelve years after Bohm proposed his theory, the situation remained at a standstill. Most physicists, in tandem with the wide-ranging intellectual conservatism of the 1950s, were not keen on changing a successful theory to respond to some metaphysical need for realism—especially if such a theory invoked nonlocality, a property most preferred to leave outside physics. Pragmatic efficiency trumped interpretational unease. They were happy to ignore that at stake was either a new theory of the fundamental properties of matter consistent with our naïve expectations that Nature is ultimately comprehensible, and thus driven deterministically by local causes, or the acceptance of the weirdness of quantum mechanics as an unmovable obstacle precluding knowledge of the ultimate essence of reality. In other words, most physicists were happy to side with Bohr and Heisenberg and declare the ultimate essence of reality—whatever that meant—unknowable.

Few seemed interested in looking for what would be Einstein's preferred choice: a hidden-variable theory that was local. Such a theory could, at least in principle, restore reality to quantum physics, exorcizing the dreaded nonlocality. Should one go ahead and spend time searching for such theories? Or was it all for naught?

The surprising answer came in 1964, when Irish particle physicist John Bell had a brilliant idea. As he later wrote, it was Bohm's theory of hidden variables that inspired him: "I saw the impossible done. It was in papers by David Bohm."[3] Bell found a way to experimentally distinguish between quantum mechanics and extensions with *local* hidden variables and decide whether the usual formalism is incomplete in the sense that Einstein and others believed it was.

Working at CERN at the time, Bell took advantage of a sabbatical leave in the United States to think about such a neglected and dangerously philosophical topic. I met Bell once, when I was beginning my graduate studies at King's College London in the early 1980s. Somewhat disillusioned with the research line my advisor proposed, I started to rekindle my interest in the foundations of quantum mechanics, which I had nursed since my undergraduate encounter with the celebrated textbook *Feynman Lectures on Physics*. Taking advantage of a conference at Oxford, I approached the famous man after his seminar on recent experimental tests of his famous inequality.

"Dr. Bell, my name is Marcelo Gleiser, and I'm working with John Taylor on supersymmetric theories."

"Good—excellent topic of research."

"Yes, but the fact is I've been interested in the foundations of quantum physics since I was an undergraduate. I even wrote to David Bohm asking if he would like to supervise my thesis work, but he said he wasn't taking students any more." (Bohm was then at nearby Birkbeck College, also in London.) Bell's eyes twinkled ever so briefly when I mentioned Bohm's name.

"Well, I find your interests laudable and rare for someone your age. However, I'd suggest that you don't do your thesis work on such topics."

"And why not?" I asked, already guessing the answer.

"You should work on something solid first, something that the community endorses. Until you have a strong reputation as a physicist, no one will listen to what you have to say about the foundations of quantum mechanics. And even then, it's not a sure shot at all, believe me."

"Okay, I understand," I replied, trying to hide my obvious disappointment. "Maybe later on in my career."

"Yes, that's what I did, at any rate."

So ended my only encounter with John Bell. I guess this book is my first attempt to wrestle with the quantum ghost, and I hope a prelude to more technical work to follow. After all, it's been thirty years; if my reputation is not secured by now, it never will be.

We have seen that the EPR thought experiment explored the relation between a particle's position and momentum in order to question the completeness of quantum mechanics. Bohm simplified EPR's prescription using a particle's spin instead. This was a clever move, since spins are not only cleaner but also easier to measure. Contrary to the position of a free particle, which can be everywhere in space (being thus a "continuous variable"), the spin only assumes a few discrete values. While a classical top, be it a toy or the revolving Earth, can spin with any speed (or better, angular velocity), quantum particles have only three possibilities: no spin at all (like the Higgs boson), integer (like the photon), and one-half multiples of the quantum unit of spin (like electrons and quarks), Planck's constant h divided by 2π, $h/2\pi$. You can't change the spin of a quantum particle; it is an indelible part of what it is, of its identity.

To simplify, let's represent the unit of quantum spin by s. (So $s = h/2\pi$.) Electrons, protons, and neutrons spin only with $s/2$, while photons spin only with s. The spin may point in any direction of space, although there are ways to influence it—for example, by using a magnetic field. Let's focus on the vertical direction perpendicular to the direction of the particle's motion and denote it the "z axis." If you orient a magnetic field in the vertical direction, electrons (or any spin-1/2 particle, in units of s) will tend to align either with it or opposite it. We usually call these "spin-up" and "spin-down" states. This simplifies things enormously, since we can then say that the particles will either have spin $+ s/2$ or $-s/2$; there are only two possible choices. To make life even simpler, let's call these two choices $+1$ and -1, for spin-up and spin-down, respectively.

In his thought experiment, Bell imagined a source that produced pairs of entangled spin-1/2 particles always with total spin zero: so if one points up (↑, represented as $+1$), the other necessarily points down ($\downarrow$, represented as -1). As in the Alice and Bob experiment, the particles flew off in opposite directions and were made to pass through detectors that could determine the direction of their spin. Call L the detector on the left and R the detector on the right, as indicated below:

$$L-----(SOURCE)-----R$$

If all pairs of electrons and the two detectors were always aligned in the vertical direction, we would obtain a perfect correlation: whenever one is measured to be up, the other will be down, and vice versa. The amazing thing here is what we already know from our discussion with polarized photons, that the entangled pair behaves as a unit, the second one always "knowing" which direction to point, although "knowing" is surely the wrong way to describe it. If, in quantum mechanics, we can only say a particle has a certain property once it is measured, Alice's particle only "became" spin-up once she measured it to be so. And how could Bob's particle know that so quickly? As Seth Lloyd wrote in his book on quantum information, it is as if twin brothers at two faraway bars and with no means of communicating always ordered the opposite of one another. If one says, "Beer," the other instantaneously says, "Whiskey"; if one says, "Whiskey," the other says, "Beer."[4]

To find a difference between quantum mechanics and possible extensions, Bell added a variant to the experimental setup.[5] We could choose to measure a particle's spin in *any* direction, not just vertical. Let's say we fix two directions: vertical as before and at 30 degrees with the vertical. Each detector could be set to measure either of these two possible directions. Call L| and R| the vertical direction for detectors L and R, and L/ and R/ the other direction. There are four possible joint orientations for the two detectors: (L|, R|), (L|, R/), (L/, R|), and (L/, R/). Since the electrons can only point either up or down along each of the two directions, the detectors can only read two possible numbers, $+1$ or -1. So once the detectors are set in a chosen direction, each measurement will return a pair of two possible numbers: $(+1, +1), (+1, -1), (-1, +1)$, or $(-1, -1)$.

Note that for the cases (L|, R|) and (L/, R/), corresponding to both detectors in the same direction, the outcomes are fixed by angular momentum conservation, a fundamental law of Nature that says that the total amount of rotation remains the same in an unperturbed physical system. If L| $= +1$, then R| $= -1$, or vice versa. If L/ $= +1$, then R/ $= -1$, or vice versa. This is the case we just discussed of perfect correlation between the two particles.

These are the four independent possible outcomes when one assumes that there is no correlation between different directions of the spin for the particles hitting detectors L and R, corresponding to the locality that Einstein and Schrödinger so wanted to see realized in Nature. The expectation is that nothing special happens for the mixed combinations (L|, R/) and (L/, R|).

Going over the four possible orientations for the two detectors for each run, the experimentalist can create a table with the results for many runs, writing down the pairs of numbers for each measurement.[6] (In other words, each run corresponds to four separate measurements, one for each arrangement of the detectors.) She could also study the values for certain relations between the pairs of readings for each run. An interesting one is the following, which we will call C:

$$C = (L| \times R|) - (L/ \times R|) + (L| \times R/) + (L/ \times R/) =$$
$$(L| - L/) \times R| + (L| + L/) \times R/$$

The last expression was obtained by rearranging the terms. The experimentalist computes C for each run, going through the four possible settings for the detectors, recording the values for the spin of both particles each time.[7] If everything goes according to local theories, the results must be as follows: since L| and L/ can only be either $+1$ or -1, one of the two terms in parentheses will always vanish, while the other will be either +2 or −2. (For example, if L| = +1 and L/ = −1, the first term will equal +2, and the second will vanish. If, instead, L| = −1 and L/ = +1, then the first term equals −2, and the second will vanish.) And since R| and R/ are either +1 or −1 for each run, the end result for C will always be either −2 or +2.

The experimentalist computes C for each run and records it. (See notes 6 and 7 for details.) Say she does it N times. She can then compute the average value for C, C_{ave}, which is $C_{ave} = (C_1 + C_2 + \ldots + C_N) / N$, where by C_1 I mean the value for C for run 1, by C_2 for run 2, and so on until the last, C_N. Since each time C can only either be −2 or +2, C_{ave} can only be a number between −2 and +2. We can then write, $-2 \leq C_{ave} \leq +2$. (For example, if, after four runs the experimenter gets $C_1 = +2$, $C_2 = -2$, $C_3 = +2$, and $C_4 = +2$, she would compute $C_{ave} = (2-2 + 2 + 2) / 4 = 1$.)

Local theories thus predict that the average value of C will always be between −2 and +2. However, when the computation for C is done using quantum mechanics, we find stronger correlations for particles oriented in different directions and thus obtain a different result: measurements of the spin of the two particles in different directions are *not* completely independent from one another. As a consequence, the value for C can go beyond −2 or +2. For certain choices of angle from the vertical, quantum correlations between the spins of the particles are *larger* than what local theories predict. In other words, for quantum mechanics the inequality $-2 \leq C_{ave} \leq +2$ *should* be violated. Bell had found an unambiguous *experimental* test to distinguish between traditional quantum mechanics and modifications with hidden variables that assume locality in the sense of EPR.

While I was writing these lines, Zeilinger's group, in an international collaboration involving the US National Institute of Standards and Technology and groups in Germany, performed a state-of-the-art experiment involving entangled photons confirming that Nature does work through spooky actions-at-a-distance.[8]

The novelty was to institute an efficient accounting for *all* the photons participating in the experiment, something that up to now had been hard to do. (Some photons always escape detection and thus are not counted.) This is crucial, since it eliminates possible biasing from the source and detectors (counting only the photons that "matter"), making the result more foolproof. Zeilinger's experiment is the latest in a long tradition that started in 1972, when John Clauser and Stuart Freedman of the University of California at Berkeley found a violation of Bell's inequality consistent with quantum mechanics. Fast-forward to the experiments by Alain Aspect and his group in the early 1980s and by Zeilinger and his group in the 1990s, and the results are rock solid: in every case, not only is an expression of Bell's inequality violated, but the violation is in excellent agreement with the prediction from quantum mechanics.

The net result of these experiments, and what's really disconcerting, is that Einstein's hope for a local theory explaining what's going on—a theory that, like his extension of Newton's gravitational theory, could exorcize the quantum ghost—has been shot down. Taken together, the experiments rule out local theories of quantum mechanics using hidden variables to explain instantaneous action-at-a-distance. Nonlocality (also sometimes called "separability")—influences acting superluminally between members of spatially separated entangled pairs—is a ghost that seems to be real. Reality is not just stranger than we suppose. It's far stranger than we *can* suppose.

If you read the preceding paragraphs and are not shocked or perplexed then you must go back and read them again. If you skipped them, proceed and be shocked. Something influencing something else far away without exchanging information in any conventional way is spooky indeed. It defies "reasonable." It adds a dimension to reality completely foreign to our everyday perception of time and space. In fact, it *does away* with time and space, since it acts instantaneously (or at least superluminally) and at any distance (at least as has been measured thus far).

What does this mean for our perception of reality? Is this something confined to the world of the very small, a fragile quantum effect lost in the large-scale dimensions of human affairs? Or does it have meaning for the way we interact with physical reality, with each other? People talk about "synchronicity," the otherworldly ability (or belief in the ability) to sense something or someone outside the time domain—instantaneously, so to speak. "I *knew* you'd be coming here today!" or "The other day I was driving with my cousin when I said, 'I love maple syrup.' Within seconds, a road stand with the sign 'Homemade maple syrup' showed up!" Are such occurrences mere coincidences? Maybe delusions of people too eager to establish connections? Or could there be some measure of large-scale entanglement that our minds can somehow sense?

This is where we tread the extremely thin line between serious science and loopy speculation. The science is solid: quantum nonlocality is not going away anytime soon. Speculations of these effects influencing the sphere of macroscopic phenomena are, at least for now, unfounded. After Bell's inequality and the experimental verification of nonlocality between entangled particles, few physicists would claim that quantum mechanics is faulty. Most of the relevant experimental loopholes have been dealt with. Disagreement begins and tempers flare when we ask how far into our reality we can push the strangeness. A decade ago it was easier to wave the whole thing away, saying that weird quantum effects are only operational at small scales far removed from our own. But recent experiments have changed this comfortable position in dramatic ways.

In April 2004, Zeilinger's group in Vienna used a pair of entangled photons to transfer a €3,000 donation to his lab from the city hall to the Bank of Austria. To do this, one of the two photons had to travel 1,450 meters (0.9 mile) along a fiberglass cable without compromising its entanglement with its partner. The year before, Zeilinger had succeeded in sending entangled photons across the

Danube River from atop two sewage towers at opposite margins. Raising the stakes, in 2007 Zeilinger sent entangled photons across the 144 kilometers (89 miles) separating the two Spanish islands of Tenerife and Las Palmas. A laser created the pair at an observatory in Las Palmas; they were then received at another in Tenerife. The point of the experiments was to demonstrate that entanglement survives for long distances in *open air*, across the hot and turbulent atmosphere. Zeilinger has plans to replicate the feat in space, using the International Space Station as the source for entangled photons. The particles would then be sent to detectors positioned far away from each other on Earth. Preliminary tests with a Japanese satellite were promising. It appears that nonlocality is more robust than most previously anticipated. This being the case, why don't we sense more of it? Or do we?

CHAPTER 27

CONSCIOUSNESS AND
THE QUANTUM WORLD

*(Wherein we discuss the role consciousness
might play in the quantum realm)*

I mentioned before my encounter with the great physicist John Bell, who advised me to stay away from research on the interpretation of quantum mechanics during the early stages of my career. I also mentioned that before I met Bell, David Bohm had told me that he was no longer taking students. Hard as I tried, the doors into the foundations of quantum mechanics were being closed one by one. In a somewhat desperate move, and already well into my PhD—having published several papers on the cosmology of unified theories with extra space dimensions—I went back to someone whose book had inspired me in my freshman year at university, even though by this time I already had misgivings about the way he tried to tie together modern physics and Eastern mysticism: Fritjof Capra. On December 7, 1984, I sent Capra a heartfelt letter exposing how my views on physics clashed with the mainstream "shut up and calculate" approach. Smitten by the romantic archetype of the scientist as rebel, I contemplated working with him on the relationship between mind and the quantum. Fortunately (from my current perspective), I was too late; even though affiliated with the Lawrence Berkeley Laboratory in Berkeley, California, Capra didn't have a permanent faculty position and wasn't advising students. No doubt my career would have taken a very different path had Capra taken me under his wing. In retrospect, I am glad he didn't.

I was twenty-five and in search of ways of connecting the rational scientific approach that I was learning in school with a strong sense of spirituality I had nurtured since an early age. At around the same time, I read Colin Wilson's *The Philosopher's Stone* and kept wondering if there was more to our brains than what we were using. Wilson's science fiction novel presciently explored how electric stimulation of the neocortex could turn our brains into dazzling genius mode.[1] Could this extra power be pent up in there, waiting to be explored? To add to the mix, a few years earlier I had been stunned, together with millions of viewers across the globe, by the trickery of the Israeli "psychic" Uri Geller and his spoon-bending powers. How on Earth could he do that? And the "broken" watches people started up by holding them and telling them to tick again? Following Geller's instructions, I brought my grandfather's old wristwatch back to life after being broken for years. At the time, the work of the fabulous magician and skeptic James Randi demonstrating how such tricks were performed with plain old illusionist moves were not as widely advertised as Geller's TV appearances. How could reason compete with the excitement of the magical?[2]

In my youthful enthusiasm, I knew I was not alone in my quest to connect physics with the beyond. Many great Victorian physicists had their mystery days, including several Nobel Prize winners: Lord Rayleigh, who had explained the color of the sky; J. J. Thomson, the discoverer of the electron; William Ramsay, discoverer of the noble gases; Sir William Crookes and Sir Oliver Lodge, both leading physicists of the age. All of them, and many more, engaged in occult activities searching for evidence of telepathy, communication with the dead, psychokinesis, and other psychic and supernatural manifestations.[3] At the time, space had just been inundated with invisible electromagnetic waves, ethereal vibrations emanating from dead and living matter. Guglielmo Marconi had perfected the transmission and reception of radio waves, teasing sounds and voices out of thin air. What else could be hovering unnoticed in this elusive realm?

New science always flirts with the boundaries of the possible. If our limited sense of awareness leaves so much unnoticed, why not much more? What if there is a soul that somehow survives the material disintegration of the body? Cutting-edge science blended with all-too-human aspirations of everlasting life to create a timeless realm where spirits coexisted. If we were to open the proper channels of communication, they might even acquiesce to our desperate calls. Crookes, Lodge, and Thomson took part in hundreds of séances, each time with renewed hope that something life-changing would happen. Not so long ago science was still malleable enough to allow for some of its great masters to engage in such pursuits. It was no coincidence that I had opted to go to England for my PhD. I too secretly wished to establish a connection between our reality and the magical, invisible reality lurking in the shadows of the possible.

What the Victorian gentlemen of science were attempting—to establish a bridge between the material and the spiritual—was revisited, even if in more formal clothing, by the gentlemen founders of quantum mechanics, as they explored the relation between quantum physics and the role of the observer. Quantum physics triggered a clash between the ordinary and the extraordinary, between the routine awareness of our everyday experiences and an alternative world where strangeness was the norm. How should we position ourselves? Should we battle the strangeness and insist, as Einstein did, that reality remained "reasonable" underneath? Or should we let go of such "old-fashioned" realist ways and embrace the new world of the quantum, taking its defiance of the normal as the new world order?

If you opted for embracing the quantum strangeness, the next question would be how far you were willing to take it. Since different interpretations of quantum mechanics are not easily amenable to experimental discrimination, most physicists prefer not to engage. Whatever you believe quantum mechanics is telling us about the world, in the end what matters is what we measure with our detectors. Let's explore the weirdness without getting bogged down by subjective interpretations. After all, isn't science supposed to work precisely because of its independence from subjective choices?

Such matter-of-fact dismissal of the intricacies of quantum physics shocks physicists from the other camp. "How can you sleep at night," they'd ask, "knowing that we have no grasp on the essence of reality? Nonlocality removes the clear spatial separation between the classical (big) and the quantum (small). To close your eyes to this is as bad as what some churchmen did to Galileo, refusing to look through his telescope."

Let me advance that there is no resolution to this impasse. Here is how Maximilian Schlosshauer, Johannes Kofler, and Anton Zeilinger summarized the situation, after having conducted a poll among attendants of a conference titled Quantum Physics and the Nature of Reality held in July 2011 in Austria:

> Quantum theory is based on a clear mathematical apparatus, has enormous significance for the natural sciences, enjoys phenomenal predictive success, and plays a critical role in modern technological developments. Yet, nearly 90 years after the theory's development, there is still no consensus in the scientific community regarding the interpretation of the theory's foundational building blocks. Our poll is an urgent reminder of this peculiar situation.[4]

There are many ways to consider this situation, ranging from the more reasonable to the more extreme. Starting with the more reasonable, the old Copenhangen Interpretation sets down the rules of the game: there is a clear separation between a quantum system and a classical measuring device. We, the observers, never interact directly with the quantum system; the detectors do. We interpret the interactions between the quantum system and the measuring device after a series of amplifications lead to a flash or a click or a tick or a track or a photographic or digital register. The wavefunction, the fundamental entity of quantum physics, is a mathematical quantity representing potentialities—possible outcomes of a measurement. It is not a physical quantity in the sense of having any kind of physical reality. Contrary to classical physics, where the equations of motion refer directly to a concrete moving object (a ball, a wave, a car), in quantum physics the equation refers to a probability amplitude. Let's say, for example, that we are interested in measuring the position of a particle. Before the measurement, the particle's wavefunction spreads through all of space (or the space where the particle can move, if constrained), reflecting the different probabilities of it being here or there. Schrödinger's equation describes how this wavefunction evolves in time, given the possible forces that act on the particle. When a measurement is made and a particle is detected in a certain position, the wavefunction "collapses": it ceases to be a potentiality and becomes a reality. It does so instantaneously, shrinking from being everywhere it can be to being localized in a small region of space. Strictly speaking, the act of measurement gives reality to what is being measured, bringing it from the netherworld of quantum potentialities to the concrete world of detection and sensorial perception. More dramatically stated: to measure is to create.

Things get complicated rather fast when we start to ask questions about this scenario. When we say, "To measure is to create" (like Pascual Jordan did), who or what is creating? Are we content in attributing creation to a mechanical measuring device? Is it the Geiger counter that kills Schrödinger's cat when it registers a particle and releases the poison? Or does observation need a conscious observer, an intelligence with the *intent* to measure and the rational ability to interpret the measurement? If an intelligent observer is needed to create reality, how can we explain that the Universe existed for billions of years without a conscious observer? Was it some kind of omnipresent and omniscient God, as George

Berkeley suggested in the eighteenth century? Is it possible that the Universe collapsed its own wavefunction when it transitioned from a quantum entity at the beginning of time to a classical expanding spacetime? If so, and if nonlocality persisted through cosmic history, could everything still be entangled somehow?

Nobel laureate Eugene Wigner, who explored in depth the importance of mathematical symmetries in quantum mechanics, addressed head-on the issue of the role of consciousness in quantum physics. "When the province of physical theory was extended to encompass microscopic phenomena . . . the concept of consciousness came to the fore again: it was not possible to formulate the laws of quantum mechanics in a fully consistent way without reference to the consciousness."[5] ("Again" here refers to René Descartes and his *Cogito ergo sum*, which recognizes thought as primary.) Wigner, like Heisenberg before him, realized that any measurement needs a mind to make sense of it. There is a continuum extending from the object detected to the detector and, finally, to the consciousness of the observer. Even if in classical physics an intelligent observer is also needed to design and make sense of a measurement, the key difference is that in the quantum world the measurement gives reality to what is being measured—without the mind there is no reality. The challenge that John Bell refers as the "central problem" of quantum mechanics is to determine where the dividing line between this real classical world "out there" and this contingent quantum world "in here" resides.

Wigner created a parable known as "Wigner's friend" to illustrate his point. Imagine that Wigner's friend, an experimental physicist, has set up an apparatus to measure the spin of an electron. It could be either up or down, having a fifty-fifty chance for each outcome before the measurement, when the electron is in a superposition state. Once she goes ahead and measures it, she will find either spin-up or spin-down; after the measurement there is no more room for probabilities or superposition. Imagine that Wigner knows about the experiment but only asks for the outcome after the measurement is made. To his friend, the outcome has been fixed, and the electron's wavefunction has collapsed into one of its two possible states. To Wigner, however, the electron remains in a superposition until he asks his friend for the result. This dual behavior, argued the real Wigner, didn't make any sense. Indeed, you would be forced to conclude that before you asked your friend for the outcome, *she* would be in a superposition between two states (a "state of suspended animation," according to Wigner), corresponding to the two possible outcomes of her measurements. That was bad enough for Schrödinger's cat. "It follows," concluded Wigner, "that the being with a consciousness must have a different role in quantum mechanics than the inanimate measuring device."[6] Even more poignantly, "consciousness enters the [quantum] theory unavoidably and unalterably."[7]

Princeton physicist John Wheeler pushed Wigner's ideas even further with his "participatory universe." He claimed that the act of measurement was more than mere observing: it determined how history would unfold from that point on (and even backwards!); it changed the universe. As soon as an experimentalist decides to measure an electron in a specific way (say, with a particle detector as opposed to a wave interferometer), "the future of the universe is changed. [She] changed it. We have to cross out that word *observer* and replace it by the new word *participator*," Wheeler argued during a conference in Oxford in 1974.[8]

Wheeler suggested a thought experiment to clarify his point. Suppose that an experimentalist sets up a source of photons, which are to pass through a double-slit obstacle as in usual quantum interference experiments. The source can be tuned so that a single photon is released at a time. After the double-slit, the experimentalist puts a screen to show the expected pattern of bright and dark interference lines. This screen, though, is mounted on wheels so that the experimentalist can move it in and out of the path the photon follows after passing through the double-slit. Behind the screen, the experimentalist mounts two detectors, each aligned with one of the

WHEELER'S DELAYED CHOICE EXPERIMENT

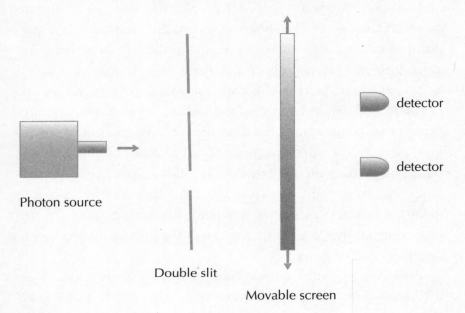

A diagram of Wheeler's Delayed Choice Experiment

slits from the two-slit obstacle. This way, if the screen is not there, the detectors can tell through which slit the photon has passed. There are two possible outcomes: either the screen is there, and the experimentalist sees an interference pattern on the screen, or if the screen is not there, she detects through which slit the photon has passed. But here is the twist: the experimentalist could decide to keep or remove the screen *after* the photon has passed through the slits.

Wheeler's point was that the photon would respond to whatever *detecting* device was chosen. He called this a "delayed choice" experiment. The experimentalist's free choice determines the physical reality of the photon (wave or particle) presumably backwards in time. This is how Wheeler explained his idea: "The past has no existence except as it is recorded in the present. . . . The universe does not 'exist out there,' independent of all acts of observation. Instead, it is in some strange sense a participatory universe." He later wrote, in an extension of his experiment to an astronomical light source, "We decide what the photon *shall have done* after it has *already* done it." Our choice interferes with the past history of the particle. Wheeler then extended this notion to the universe as a whole: the observer "gives the world the power to come into being, through the very act of giving meaning to that world; in brief, 'No consciousness; no communicating community to establish meaning? Then no world! . . . The universe gives birth to consciousness, and consciousness gives meaning to the universe.'"[9]

Even if apparently far-fetched, versions of Wheeler's delayed choice have recently been confirmed experimentally, at least for quantum systems. In 2007, Vincent Jacques and colleagues (a group that included Alain Aspect, who, the reader will remember, played a key role in demonstrating the violation of Bell's inequalities), followed Wheeler's prescription very closely, making sure that no communication at or below the speed of light was possible, that is, that the photon had no way of "knowing" how it would be measured.[10] At the end of their manuscript, the experimentalists quote Wheeler: "We have a strange inversion of the normal order of time. We, now, by moving the mirror in or out have an unavoidable effect on what we have a right to say about the already past history of that photon."[11]

Nonlocality startles us, forcing a revision of deeply ingrained concepts such as causality. Could the present really determine the past? Can such bizarre notions be extrapolated from quantum systems and their fragile existence to larger objects, and even to the Universe as whole, as Wheeler had wanted? "Is the term 'big bang' merely a shorthand way to describe the cumulative consequence of billions upon billions of elementary acts of observer-participancy reaching back into the past?"[12] Wheeler is careful to distance consciousness from the act of observation, which he understood as a registration of some sort. Meaning, understood as how consciousness will interpret that registration, is a "separate part of the story." He is ambiguous because he doesn't know the answer. No one does.

> Are billions upon billions of acts of observer-participancy the foundation of everything? We are about as far as we can be today from knowing enough about the deeper machinery of the universe to answer this question. Increasing knowledge about detail has brought an increasing ignorance about the plan. The very fact that we can ask such a strange question shows how uncertain we are about the deeper foundations of the quantum and its ultimate implications.[13]

It's no wonder that most physicists, faced with such weirdness, balk at trying to interpret quantum physics, opting to accept the Copenhagen Interpretation and moving on. Until there is a clear experimental test, interpretation is a personal choice. Another approach, no less weird but one that attracts a surprisingly large number of physicists, is known as the Many Worlds Interpretation (MWI). First suggested by Schrödinger in a lecture in Dublin in 1952 as an idea that might "seem lunatic," and then proposed in more detail in physicist Hugh Everett's 1957 PhD thesis (Everett was Wheeler's student) and expanded by Bryce DeWitt in the 1960s and 1970s, the MWI is truly radical. It asserts that there is no such thing as a wavefunction collapse upon measurement: all possible outcomes of a measurement—all potentialities—are in effect realized at once, each in a parallel world (or universe). According to the MWI, all possible histories coexist in a kind of multiverse and branch out every time a measurement is made. Schrödinger's cat is both dead and alive, in two separate universes; the electron is both spin-up and spin-down in two separate universes; the photon is a wave here but a particle there. By creating the possibility of a multiverse of outcomes—a nondenumerably infinite number of worlds—the MWI attempts to do away with the paradoxes of quantum mechanics.

Like in Jorge Luis Borges's *The Garden of Forking Paths*, where a labyrinth exists in time (and not in space, as in a usual labyrinth) such that at each fork history follows two separate alternatives, in the MWI different possible histories coexist side by side, even if inaccessible to one another. The core supposition is that the wavefunction is not simply a mathematical artifact but a real entity guiding the creation of parallel histories. The MWI is an attempt to restore reality to physics, albeit at the price of invoking the existence of an ever-growing multiverse branching out in a multitude of histories. Quantum information theorist David Deutsch, from Oxford, is perhaps its most vocal proponent. In his recent book *The Beginning of Infinity*, Deutsch doesn't mince words, criticizing the Copenhagen Interpretation as "bad philosophy," the kind that "is not merely false but actively prevents growth of other knowledge."[14] He goes on: "There the idea is that quantum physics defies the very foundations of reason: particles have mutually exclusive attributes [being particle and waves], period. And it dismisses criticisms of the idea as invalid because they constitute attempts to use 'classical language' outside its proper domain." Instead, Deutsch claims, all of the "vagueness" springing from nonlocal wavefunction collapse and observer-determined reality washes out if you embrace the reality of the wavefunction and its many-world multiverse. To most physicists, however, the choice is not so obvious.

No one could (or should) claim with any measure of confidence that the MWI has solved the measurement problem in quantum mechanics. As with Bohm's nonlocal, hidden-variable theory, it offers a strange alternative to the quandary of the wavefunction collapse by not collapsing it and introducing another level of complexity—the parallel existence of countless branching worlds, each separate and without interaction with ours. Where are such worlds that exist but which we cannot have an awareness of? When in the process of measurement does the branching occur? As of yet, there is no conclusive experiment that could distinguish between Bohm's formulation and MWI, or validate the MWI viewpoint as a viable alternative to the Copenhagen Interpretation, although some physicists argue that once interference with large objects (back to Schrödinger's cat) is doable, different histories may couple to one another, even if weakly. Until a concrete and doable experimental test comes forth, the existence of parallel, noninteracting universes offers as much guidance to the fundamental question of measurement and the nature of reality as proposing a multiverse of eternally branching universes in cosmology offers to explaining why we live in our Universe. (See Part 1.)

A major advance, albeit also controversial when presented as a solution to the measurement problem, is quantum decoherence, whereby the trouble-making quantum interference between different possible outcomes of an experiment is destroyed ("decohered") by interactions between the quantum system and the environment around it. In this view, classicality is a consequence of the loss of quantum interference: the classical world of large, nonquantum objects emerges once interactions with the environment are factored in. Some physicists present decoherence as a natural continuation of the Copenhagen Interpretation, given that the measurement process surely destroys any quantum coherence in the wavefunction. Others present decoherence as a continuation of Everett's MWI, whereby decoherence causes the splitting of worlds or histories. A variation of decoherence known as "decoherent histories formalism" or "consistent histories," originally proposed by Robert Griffiths in 1984 and independently by Roland Omnès, was rediscovered and applied to quantum cosmology by Murray Gell-Mann and James Hartle in 1990, in which one attempts to describe the Universe as a whole as a quantum system. The challenge there is that since the Universe is considered a "closed system," there are no outside observers or an environment to decohere its global wavefunction. The transition from a quantum universe to a classical one comes through its own time evolution, as different possible histories, each with a different set of probabilities, unfold separately, decohering from the whole, so to speak. The specific decohering is due to the particular events (e.g., interactions between particles) that happen in a specific history. We live in one branch of these unfolding histories, the one that has the Universe with the properties we measure. Unfortunately, it remains unclear what sort of mechanism would favor our Universe, if any at all.

Decoherence encompasses the traditional Copenhagen Interpretation view that measurement causes wavefunction collapse. A measurement is an event that enforces a violent kind of decoherence, an approximation where decoherence is idealized as exact and instantaneous. There are other kinds of "measurement" that are not so drastic but that also affect the evolution of the wavefunction. As physicist James Hartle wrote, "Probabilities may be assigned to alternative positions of the moon and to alternative values of density fluctuations near the big bang . . . whether or not [these effects] were participants in a measurement situation and certainly whether or not there was an observer registering their values."[15] In other words, the very conditions of the early universe determine its future unfolding histories, including our own emergence as an inevitable consequence of the interplay between such conditions and the intrinsic randomness of quantum physics. In this view, participants don't affect the past history of the Universe.

The decoherence approach does make clear the artificiality of the separation between a classical observer or detector and a quantum system. It further shows that the classical world of our senses is an emergent property of matter, a consequence of how quantum systems with many components interact with themselves and other systems around them. The larger the system, the larger the number of quantum mechanical wavefunctions to describe all of its components, and the more difficult it is to herd them all into coherent states that display quantum superposition. Systems in quantum superposition are extremely fragile and collapse under the minutest interference from the outside, be it a photon from the Sun, a cosmic ray, or the gravitational disturbance from a truck passing by. Decoherence helps us understand why a classical world is emergent from a quantum realm existing beyond our perception, although it doesn't really specify where the boundary between quantum and classical resides. As John Bell put it:

> The 'Problem' [of quantum mechanics] then is this: how exactly is the world to be divided into speakable apparatus . . . that we can talk about . . . and unspeakable quantum system that we can not talk about? How many electrons, or atoms, or molecules, make an 'apparatus'? The mathematics of the ordinary theory requires such a division, but says nothing about how it is to be made.[16]

Most importantly, decoherence doesn't solve the measurement problem for the simple reason that the outcomes of measurements continue to be random and not predetermined by some underlying hidden order. (For example, we can't foretell which polarization, vertical or horizontal, the photon will have before we first measure it.) Despite the added clarification decoherence brought about—we needn't ponder whether Schrödinger's cat is dead or alive or whether the Moon is not there when we aren't looking—we must still wrestle with the quantum ghost of nonlocality and with our inability to make sense of an underlying physical reality. We also don't understand the role of consciousness in determining that reality, if indeed there is one.

CHAPTER 28

BACK TO THE BEGINNING

(Wherein we attempt to make sense of what the
quantum enigma is telling us)

Quantum mechanics forces us to confront the unknowable head-on. This is why it makes most physicists uncomfortable. "Unknowable" is anathema to the core goal of the scientific program, which is designed to deal with, and ultimately make sense of, unknowns. Einstein, Schrödinger, and the scientific realists shun the possibility that Nature would keep secrets beyond our grasp. They admitted that our knowledge of Nature was limited and at best incomplete, but they would argue that the limitation was due to our own inefficiency and not to some deep, inscrutable reason. They hoped that the probabilistic nature of quantum systems was not fundamental but operational. After all, probability is essential to another successful theory—the statistical mechanics describing the behavior of gasses and systems with many particles—and there it merely reflects the practical inability of tracking the behavior of each individual particle among trillions upon trillions of them. Instead, we describe the collective average behavior of the particles and treat departures from this average behavior statistically. The realist gang expected something similar to happen in quantum mechanics, so that the probabilistic behavior was not an inherent property of small systems but an effective one, a product of our limited understanding of the true nature of the very small.

We identify a similar expectation when certain physicists pronounce that we know how to explain the origin of the Universe using quantum mechanics and general relativity. Of course we don't, and all we have thus far are very simplistic models based on a host of unproved assumptions. The expectation is not just hopelessly naïve but also philosophically misplaced, given that any model in the physical sciences is built on a scaffolding of idealized concepts, such as space, time, energy, and conservation laws. The origin of the Universe encompasses the origin of all such concepts. And where do they come from? Furthermore, the models are formulated with what we call "boundary conditions," which presuppose a clear separation between the object of study and its surroundings. Clearly, such separation is ill defined when the object of study is the Universe as a whole, even when we consider the peculiarities of curved geometries.

When we try to explain the origin of the Universe with physical models, the best we can hope for is to construct a viable description of how the first moments of cosmic history unfolded in ways consistent with the data we can gather. This is a vast and tremendously exciting undertaking, but it is not the same as an explanation for the origin of all things. To explain the origin of all things we would need to start by explaining the origin of the physical laws that describe this Universe, something that is beyond the jurisdiction of current physical theories, including those that ascertain the existence of a multiverse where laws can vary. As we discussed in Part 1, claims of varying laws of Nature in different cosmoids are muddled at best. Moreover, if we want to make any kind of progress in understanding the quantum origin of the Universe, we need to further clarify the meaning of nonlocality in quantum theory—which brings us back to our discussion of entanglement and decoherence.

Fortunately, four decades of amazing experiments have taught us a few things. We know that extensions of quantum theory using local hidden variables have been ruled out: if there are hidden-variable extensions of the quantum theory, they must be nonlocal and won't alleviate Einstein's anxiety with spooky actions-at-a-distance. Nonlocality is an indelible feature of entanglement, and entanglement is an indelible feature of quantum mechanics. The reason we find it so unfamiliar is because it is. Entangled states are fragile and hard to maintain for long periods of time and for large distances; experimental physicists go through all sorts of trouble trying to nurse entanglement into longevity. Many different kinds of interactions with the environment—thermal, vibrational, gravitational, and even between the object's own jiggling atoms—act to destroy them. The reason why the Moon is not at many places at once along its orbit is because the Moon is not an isolated system: photons from the Sun hit it all the time (that's why we can see it), as do cosmic rays; there are vibrations of the myriad atoms making it up, and gravitational forces from the Sun, the Earth, you, and so forth. All contribute to destroy possible superpositions of "Moon-here" and "Moon-there" states. Large objects are hard to isolate from the decohering influences of the outside world. Our classical reality is etched out from the decohered shadows of the quantum realm.

The boundary between the classical and the quantum is fluid. Some systems can display typical quantum behavior for surprisingly large distances and lengths of time. Zeilinger identified entangled photons across hundreds of miles; crystals and large molecules can be coaxed into superposition and display interference patterns as photons and electrons would. With proper care, entanglement can survive and be protected. However, we must remember that this happens in the highly artificial world of laboratories, through the diligent work of creative experimentalists. There is no question in my mind that such achievements will become increasingly refined and that they will, in the coming decades, be applied in the construction of the first working quantum computers and that quantum cryptography will become more widespread, as will other uses of entanglement and the essential randomness of quantum systems.

These applications, exploring some of the strange properties of the quantum world, raise an interesting question: Can we elevate superposition and entanglement to the level of macroscopic objects, possibly even living systems? Is it all a matter of funding, as Zeilinger once declared, or are there more fundamental obstacles to extending quantum effects to increasingly more complex systems? What would it mean to create a quantum superposition of a living bacterium as it passes through double slits? Can life survive quantum interference?[1] This is, perhaps, a different way of formulating Bell's key "Problem" of quantum mechanics, the discontinuity between the two realms. Decoherence may explain why we perceive a classical reality that is so profoundly different from the quantum realm. But could we fabricate quantum effects and amplify them to levels that become part of our reality? In other words, if the essence of reality is quantum, can we make it not just clicks and flashes in classical detectors but part of our direct awareness? And if so, could we then catch a glimpse of its true meaning?

I don't think anyone knows the answer to this question. And I believe we can't. The essential reason is not one of experimental limitations but one that relates to what we have learned thus far from quantum physics. The fundamental discovery from EPR-related experiments is that randomness is an inherent aspect of Nature. When Alice or Bob measures the spin or polarization of their entangled particles they can't know which result, "up" or "down," they will get. We don't have a theory that predicts the outcome of a single quantum measurement. Worse, given that local hidden variables have been ruled out, this theory doesn't seem to be possible, even in principle. So if "a glimpse of its true meaning" reflects the old realist expectation of complete knowledge of reality, then we are out of luck. Our approach to knowledge needs to be redefined in light of the discoveries of quantum mechanics. There are aspects of reality that are permanently beyond our reach. The

Island of Knowledge must exist as an island, surrounded not only by what we don't know but by the unknowable nature of the quantum realm.

There is nothing defeatist about this realization. The goal of science is to clarify, to the best of our ability, the way Nature works. Science is not supposed to answer all questions. In fact, this kind of expectation is meaningless, especially when confronted with the nature of knowledge as we have addressed it in this book: ever-exploratory, ever-changing, reflecting so clearly how we approach the world, the kinds of questions we ask and *can* ask at a given time. The knowledge that we have defines the knowledge that we can have. Still, this is just what a physicist might call the initial condition: a few steps from the beginning, the game is open-ended and unpredictable. As knowledge shifts, we ask new kinds of questions that couldn't have been anticipated.

We now know that we must accept nonlocality as part of physical reality, that there are long-range quantum effects that seem to exist outside the boundaries of space and time. The new frontier that is opening up will continue to explore different features of entanglement, including how to extend its applicability to larger systems and in more adverse environments. Inevitably, we will have to confront the role of quantum effects in the brain and its possible relevance to consciousness, beyond what Wigner proposed. Could Wheeler's participatory universe have implications for the Universe as a whole? Information seems to be a key player in determining the physical nature of quantum objects. How we set up the experiments and how we choose to ask questions determine what they become when detected: if we have no information about the path taken, interference takes place; if we do, interference is absent. At least when it comes to quantum systems, reality depends on how we interact with it.

Here we encounter the question of *intent*, of how we choose to interact with reality. The detectors may be the ones that detect and "collapse" the wavefunction, but we are the ones to set them up. Without an interpreter with consciousness of a certain complexity, reality is not even a question. And in our case at least, this conscious "interpreter" is in our brains. It is thus natural to ask whether our brains are classical or quantum devices. Or, more scientifically, to what extent, if any, quantum effects are relevant to how the brain functions.

Although ideas of Roger Penrose and Stuart Hameroff, such as exploring quantum coherence in microtubules, seem to have been discredited by experiments and theoretical calculations,[2] the complexity of the subject is such, and our current knowledge so primitive, that much remains open. There may be a role for quantum effects at the level of intersynaptic gaps, for example, as ions traveling from one synaptic end to another diffract as they pass through acceptance gates, or something yet to be proposed. Just as in the case of photosynthesis, in which quantum effects seem to play a major role in optimizing and accelerating the search process for the best energy pathways, there may be equivalent effects in the brain that could have an impact on its information processing efficiency and thus on the existence of different levels of consciousness. Even if current suggestions on the possible role of quantum effects in consciousness seem implausible, it is fair to say that the issue remains wide open.

We are transitioning from the atomic age to the information age. Our metaphors, predicated as they are on our knowledge, are changing from those of the Cold War terror of mutual-assured destruction of the 1960s and 1970s to a world where cultural barriers are being crossed with unprecedented ease, a world where the same products and technologies are available to an ever-growing sector of the population. New disciplines are emerging to reflect the new knowledge and its potential: quantum information and quantum computing, network theory, data mining and its applications, complexity theory, and so forth. At their core is the notion that information is the key to knowledge. We must thus explore the notion of information and how it interacts with and defines knowledge. Not surprisingly, we will find that there are strict limits on how much information we can extract from the world, from mathematics to computer science. More surprisingly, such limits have much to say about who we are and how we find meaning in an age of science.

PART III

MIND AND MEANING

A mathematician, like a painter or a poet, is a maker of patterns. If his patterns are more permanent than theirs, it is because they are made of Ideas.
    —G. H. HARDY, "A MATHEMATICIAN'S APOLOGY"

Then here in Copenhagen in those three years in the mid-twenties we discover that there is no precisely determinable universe. That the universe exists only as a series of approximations only within the limits determined by our relationship with it. Only through the understanding lodged inside the human head.
    —MICHAEL FRAYN, *COPENHAGEN*

We now understand that the human mind is fundamentally not a logic engine but an analogy engine, a guessing engine, an esthetics-driven engine, a self-correcting engine.
    —DOUGLAS R. HOFSTADTER,
    IN HIS FOREWORD TO *GÖDEL'S PROOF*

These two ways of thinking, the way of time and history and the way of eternity and of timelessness, are both part of man's effort to comprehend the world in which he lives. Neither is comprehended in the other nor reducible to it. They are, as we have learned to say in physics, complementary views, each supplementing the other, neither telling the whole story.
    —J. ROBERT OPPENHEIMER,
    SCIENCE AND THE COMMON UNDERSTANDING

CHAPTER 29

ON THE LAWS OF HUMANS
AND THE LAWS OF NATURE

*(Wherein we discuss whether mathematics is an invention
or a discovery and why it matters)*

We humans share a compulsion: to make sense of the world and to figure out how we fit in, individually and collectively. That we have remained as active in these pursuits as we have for thousands of years shows that in our essence we are not so different from our distant ancestors. The methods and questions change, but the need to know, the urge to make sense of life, is the same.

As people noticed the preponderance of rhythmic patterns in the skies and on land, it was natural to suppose that behind the plurality of motions and forms there was an ordering trend, as if reality were the handiwork of invisible purveyors of regularity. The nature of these purveyors is a central question of religion and of science. As such, it is also a central question of this book. Is it gods or laws or both or neither that bring order to the world? Throughout history, this question has spawned hundreds of creation myths, sacred narratives that attempt to make sense of the origin and nature of all things. Irrespective of the where and the when of a given mythic narrative, the origin of things is always associated with the emergence of order, with or without a divine hand to guide it.

The display of order and regularity in Nature—day and night, seasons and tides, a Moon with phases, planets that return, the life and death cycle of plants and animals, gestation periods—requires a methodic counting and organizing as a means to gain some level of control over what is otherwise distant and unapproachable, the trends of a world evolving in ways clearly beyond human power. How else would pattern-seeking humans order their sense of reality if not through a language capable of describing these patterns, of analyzing them, of exploring their repetition as a learning tool? The mathematization of Nature, and the ordering of observed trends in terms of laws, is one of the distinctive achievements of our species. However, since most people find laws in the social realm more familiar, it is instructive to first discuss the differences between the laws of Nature and the laws of humans.

While the laws of humans seek to order and control individual and social behavior so as to make communal life less risky, the laws of Nature are deduced from long-term observation of a wide variety of phenomena. While human laws may vary from culture to culture and time to time, based as they are on moral values that lack universal standards, the laws of Nature aim at universality, at uncovering behaviors that are true—in the sense of being verifiable—across time and space. Thus, while one group may find certain rituals acceptable and others barbaric (such as female circumcision), stars across the Universe have been burning according to the same rules since they've first appeared some two hundred million years after the Big Bang. Likewise, while in some countries the death penalty is abhorrent and in others it is exercised with fanatical zest, molecules across trillions of planets and moons in this and other galaxies combine and recombine in chemical reactions that follow patterns based on well-defined laws of conservation and of attraction and repulsion.[1]

The variations in the laws of humans show that we know little of ourselves and of what are, or should be, truly universal moral standards. On the other hand, the sense of trust and finality of the

natural laws, their apparent certainty, has inspired many a movement to use them as a basis for all laws, including human laws. The
Enlightenment, of course, is a well-known example. But this trend
existed well before the eighteenth century. Take Plato and his Ideal
Forms. We find a sense of awe with the power of mathematics and
even more with the power of the human mind to have conceived of
such a gateway to eternal truths. To a large measure, Plato caught
this fever from the Pythagoreans, who had elevated mathematics to
godlike status: with mathematics, humans transcend their mortal
nature to join God's timeless mind.

The power of mathematics comes from its being detached from
physical reality, from the abstract treatment of its quantities and
concepts. It starts in the outside world, the world as it is perceived
by our senses, when we identify approximately circular and triangular forms in Nature, or learn how to count and measure distances
and time. But then mathematics takes a simplifying step and lifts
these asymmetric shapes from Nature and idealizes them as symmetric, so that we can more easily construct mental relations with
them. These relations and their progeny may or not be applicable
back to the study of Nature. If they are, they may be used in a scientific model of some kind. If not, they may remain forever locked
in the abstract realm of ideas they inhabit. This transplanting of
forms and numbers from Nature, which allows for the abstract manipulation of number and form, is also why mathematics is always
an approximation to reality and never reality as it is.

Still today, Platonists count this detachment as a blessing, as
the only way they can peruse higher meanings in pursuit of eternal
truths. To them, it is a cleansing from the impure ugliness of an
imperfect world. They would claim that this abstract, mathematical
reality *is* reality and that mathematics is the only way to access it,
somewhat like plucking fruit from the mythic Tree of Knowledge.
(With the added bonus of not facing the Fall.) In the words of the
great mathematician G. H. Hardy, "I believe that mathematical reality lies outside us, that our function is to discover or observe it,

and that the theorems which we prove, and which we describe gran-
diloquently as our 'creations', are simply our notes of our observations."[2] Or, even more dramatically, "'Imaginary' universes are so
much more beautiful than this stupidly constructed 'real' one; and
most of the finest products of an applied mathematician's fancy
must be rejected, as soon as they have been created, for the brutal
but sufficient reason that they do not fit the facts."[3]

Others see this romantic view of mathematics as a form of
cryptoreligion, a romantic belief system that has little to do with
reality. To them, mathematics is very much the product of how
our brain functions and perceives reality in an inseparable alliance
with the body: the way we think depends on our bodies, on the way
our bodies evolved to become what they are. As George Lakoff
and Rafael E. Núñez wrote in the Preface of their thorough study
on the roots of mathematical thinking, *Where Mathematics Comes
From*,

> Human mathematics, the only kind of mathematics that human
> beings know, cannot be a subspecies of an abstract, transcendent
> mathematics.
>
>   Instead, it appears that mathematics as we know it arises from
> the nature of our brains and our embodied experience. As a con-
> sequence, every part of the romance appears to be false.[4]

Indeed, belief in the existence of a mathematical realm wherein
lies an infinitude of true statements that the human mind can pluck
with greater or lesser efficiency, given the imagination and ability
of the plucker, has all the ingredients of a religious fantasy: an
imaginary world existing in a reality parallel to our own, where
deep truths are hidden; eternal truths to which only a few chosen
ones have access through their ability to see, like prophets, what
others can't; truths that only those who are able to apprehend their
meaning can translate to the enlightenment and wisdom of common folk.

Mathematician Gregory Chaitin, who played a key role in extending Gödel's and Turing's incompleteness results into the world of algorithmic information theory (more on this later), expressed his belief in the reality of the Platonic realm in an interview: "I like to have the fantasy that I haven't thrown my life away completely and I didn't just invent [my results] but expressed some kind of fundamental reality out there."[5] At the end of the interview, however, he confesses that his lifelong research in complexity theory had forced him to acknowledge the experimental (or invented) side of mathematics, even if he still opts for a middle way from a philosophical standpoint.

Others, like the leading British mathematician Sir Michael Atiyah, agree, claiming that timeless truths exist, a "background fundamentally there to be discovered," but that the individual personalizes it, giving it their own "imprint, illuminating it" with a unique light.[6] A brave attempt, but further reflection shows that Atiyah's position is not really a middle way, since it accepts the existence of a mythic mathematical realm.[7]

I find this sort of belief wholly unfounded. Einstein agrees. In his essay "Remarks on Bertrand Russell's Theory of Knowledge," he declared: "The series of integers is obviously an invention of the human mind, a self-created tool which simplifies the ordering of certain sensory experiences."[8] Musings about a Platonic realm of eternal mathematical truths may serve as a guiding inspiration for mathematicians but have only as much substance as Paradise to a Christian believer: "It exists if I believe it does, and the power of my conviction is all I need to feed from." There is no proof that such transcendent truths exist beyond human perception. Isn't it more plausible to simply state that the human mind has the remarkable ability to create and manipulate abstract concepts with its power of logical reasoning and embodied cognition without having to attribute some sort of intangible reality to it?

Astrophysicist Mario Livio, in his book *Is God a Mathematician?*, provides an engaging history of the discovery versus invention debate, visiting the thought and work of some of the greatest mathematicians of all time. He attempts to conclude that there isn't a simple answer: "Typically, the concepts were inventions. Prime numbers as a concept were an invention, but theorems about prime numbers were discoveries."[9] One problem with this line of reasoning, of course, is that we cannot be certain of what is a discovery without a road map to the elusive Discoveryland. But in the end, Livio moves in the direction of the cognitive scientists, acknowledging the essential role of human neurological makeup to explain the effectiveness and uniformity of mathematics.

An intelligence capable of counting and of the notion of infinity can come up with the basis of arithmetic and even set theory. It is known that certain animals, such as chimpanzees and crows, can count to a few digits. But they stop there, unable to conceive of large numbers and, more to the point, that counting never ends. As Lakoff and Núñez point out, only a complex mind can contemplate the notion of infinity, the jump happening when infinity is seen not just as "potential" (counting or drawing a line without stopping) but as "actual," that is, infinity as a thing in itself. We can't count to it, but we can hold it in our heads.

Nobel-laureate physicist Eugene Wigner, in his essay "The Unreasonable Effectiveness of Mathematics in the Natural Sciences," brought the usefulness of mathematics in the physical sciences into focus, claiming it to be an unexplainable gift: "The enormous usefulness of mathematics in the natural sciences is something bordering on the mysterious and . . . there is no rational explanation for it." A pioneer of the application of a branch of mathematics called "group theory" to quantum mechanics, Wigner shared his perplexity at how so many results in physics used mathematics that was not created for that purpose, or for any purpose at all: "The miracle of the appropriateness of the language of mathematics for the formulation of the laws of physics is a wonderful gift which we neither understand nor deserve."[10]

There is indeed a beautiful complementarity between the work of mathematicians and that of theoretical physicists, as mathematics is continually applied to physical problems with uncanny success. However, Wigner's perplexity, shared by many physicists, is not justified. First, as the great mathematician G. H. Hardy happily recognized, "The geometer offers the physicist a whole set of maps from which to choose. One map, perhaps, will fit the facts better than the others, and then the geometry which provides that particular map will be the geometry most important for applied mathematics."[11] Many mathematical ideas have absolutely nothing to do with physical reality. Physicists will pick those that are useful for them to accomplish their goals. As any theoretical physicist knows very well, the absolute majority of the mathematical models we develop lead to nothing even closely related to a real-world application. Despite our intuition, most equations we solve are just that—equations we solve. Figuring out Nature is much harder than solving model equations.

Second, even the most abstract mathematics takes off from perceived reality. Numbers, sets, geometries: these are concepts that the human brain is wired to identify in the world. We count; we collect objects in sets (so many lions here, so many zebras there); we recognize patterns in the world around us. As Lakoff and Núñez argue, to understand where mathematics comes from we must clarify its "embodiment," that is, how our thought processes are the result of our cognitive makeup. Third, the notion that "truth is beauty and beauty is truth"—that there is an aesthetics of beauty in mathematics mirrored in its applications to Nature—is fallacious. Surely there are many beautiful symmetries in Nature, patterns that repeat themselves such as spirals in galaxies and hurricanes or spheres in soap bubbles and planets. There are also more abstract symmetries found in the ways fundamental particles of matter interact with one another. But most symmetries are only approximate, and many objects are not symmetric at all. Nature's creative power often hides behind asymmetries and not symmetries, as I proposed in *A Tear at the Edge of Creation.* Benoît Mandelbrot, the inventor of fractals, wrote: "Clouds are not spheres, mountains are not cones, coastlines are not circles, and bark is not smooth, nor does lightning travel in a straight line."[12] The richness is found not in isolating order above everything else, but in contrasting order and disorder, symmetry and asymmetry, as complementary players in the ways we describe Nature.

Complicating the discussion are the many examples in which the imposition of a mathematical symmetry or of a consistency condition triggered an unexpected advance in physics. Take, for example, Dirac's relativistic version of quantum mechanics, which led to the discovery of antiparticles. In attempting to build a formulation of quantum mechanics consistent with Einstein's special theory of relativity and the electron's spin, Dirac obtained not one but two solutions to his equation. One described the electron, while the other a particle just like it but with opposite electric charge. (There are a few other subtle differences, which are less important here.) After thinking that the positively charged particle was the proton (but puzzled by the huge mass difference between electron and proton), Dirac soon realized that it was in fact a new particle, an "anti-electron." In 1932, Carl Anderson detected the anti-electron experimentally, which he called "positron." More dramatically, the mathematical union between quantum mechanics and special relativity imposed the existence of a whole new class of particles, those belonging to antimatter: every particle of matter has a partner particle of antimatter.

Dirac's equation opened a window to a new world, populated jointly by matter and antimatter. Amazingly, this world turned out to be the world we live in, but with a twist: even here perfection is not realized. According to Dirac's equation and its interpretation, matter and antimatter should exist on equal footing. The mystery is that the world around us is made only of matter. This asymmetry of Nature, the excess of matter over antimatter, remains an open question, despite decades of attempts to come to a solution. And it is key to our existence and to that of everything else: because of the uncanny efficiency with which particles of matter and of antimatter annihilate themselves into radiation upon colliding with one another, a universe that began with equal amounts of matter and antimatter would quickly evolve into a void filled mostly with radiation, a world nothing like what we see around us.[13]

This and other examples have led many to believe that mathematics is more than a mere descriptive tool in physics, that physicists uncover some deep mathematical structure of the natural world, perhaps *the* deep mathematical structure of the natural world, the Platonist realm of pure mathematics transferred to physical reality. Currently, such beliefs reach their climax in the Eldorado-like Theory of Everything, the attempt to formulate a single, all-encompassing description of the material world based on the fundamental forces between elementary particles. However tempting this project may sound, giving as it does a taste of godlike truth to the discoveries of the physical sciences, a quick look at the history of physical theories teaches us otherwise. For contrary to the permanent status of mathematical results (the Pythagorean theorem will not change as we understand triangles better), physical theories are and should be always in a state of flux. Take gravity, for example. As we have seen in this book, for Newton, it was something quite different from what it was for Aristotle; for Einstein, it was quite different from what it was for Newton. We are currently going through a post-Einsteinian rethinking of what gravity is, with some physicists even questioning whether it is a fundamental force like electromagnetism or something quite different altogether.

It would be naïve and quite wrong, especially for a theoretical physicist, to claim that there is no role for mathematics in Nature. Of course it has a central and essential role, and we see this expressed in our physical theories, which are all mathematical. Symmetry plays a key role in the implementation of these theories and their applied models, given that they are excellent approximations to what we are attempting to describe. The danger, and the origin of the Platonist fallacy, is to believe that the symmetries are an imprint of Nature instead of an explanatory device we conceived to describe what we see and measure. There is a very productive alliance between the human brain and its mathematical attempts to make sense of reality. However, to claim that this mathematical framework is somehow a signature of a grand natural design that some of us can on occasion tap into is equivalent to elevating our mathematical models of Nature to a series of mystical messages from above.

If mathematical results are not snapshots of some transcendent truth but a very human invention, and if our quest for a final theory of Nature based on a unique mathematical structure is misled, why bother? Why embark on this quest if it won't take us closer to the Truth? This is the question I am often asked, mixed with accusations of being defeatist or just frustrated with my own limitations in advancing the grand quest. I imagine that more than a few readers are wondering the same thing. I find myself in the difficult role of being a romantic having to kill the dreams of other romantics. But it is time for science to be appreciated and presented for what it is, and science is *not* a God-given gift to humanity. The nexus of our quest for knowledge is not to be found outside of us but within us. Theorems in abstract mathematics, even if apparently completely disconnected from immediate reality, are the product of logical rules and concepts constructed with our minds. As Lakoff and Núñez explain, our minds function in specific ways that reflect the embodiment of cognitive tools, which facilitate the development of abstract conceptual tools. We create the mind games of pure mathematics in the convolutions of our neocortex. And our neocortex is the result of eons of evolution driven by the pressures of natural selection and genetic variability, where the link between creature and environment is essential.

It may be true that $2 + 2 = 4$ is a universal result (for any species that can count and add), but this doesn't make it less human. If other intelligent aliens find the same result (no doubt representing it by a very different set of symbols), that speaks more of the way intelligence works than of universal truths in the Book of Nature. It is not that Nature is such that $2 + 12 = 14$ or, to be fancy, that $e^{ix} = cos\ x + i\ sin\ x$. It is that human intelligence is such that $2 + 12 = 14$ and $e^{ix} = cos\ x + i\ sin\ x$, and such relations can be used to approximate and/or describe elements of physical reality such as groups of zebras or relationships between complex exponentials and trigonometric functions, useful in countless applications in all the sciences or just as abstract thought constructions.

The discussion of mathematics being an invention or a discovery, like the discussion of the nature of physical reality, points more to the importance of the human brain as a rare and wondrous oddity in the Universe than to elusive truths written in some imponderable abstract realm. The cause for celebration is not "out there" or "up above" or in the "mind of God" but in this small mass we humans carry within our cranial cavity.

CHAPTER 30

INCOMPLETENESS

(Wherein we briefly explore Gödel's and Turing's disconcerting
but all-important findings)

Taken together, the discoveries of early-twentieth-century physics went very much against the prevalent—and essentially Newtonian—notion at the time: that Nature was ultimately rational and independent of human interference or scrutiny. First, Einstein's theory of special relativity presumably imposed the need for an observer's point of view in order to interpret measurements of position and time. Then, Heisenberg's uncertainty principle seemed to have amalgamated the presence of the observer with our interpretation of physical reality in a kind of indissoluble wholeness. Full of mischief, the new physics brought back the human factor into a science that had prided itself on its rigor and absence of subjective meddling. As we have seen, there are subtleties with this statement, given that Einstein's relativity is really a theory of absolutes (the laws of Nature and the value of the speed of light being the same for all observers), and Heisenberg's uncertainties fade away as we transition from the atomic and molecular scale into the larger objects of our everyday life. Still, there is something going on here, a different way of thinking about physics and how the human factor plays a role.

Kurt Gödel's surprising and brilliant result brought a similar sort of human focus into mathematics. In 1930, at the age of twenty-three, the Austrian logician presented his two related incompleteness theorems, in which he proved, in essence, that mathematics, or more precisely, any formal system adequate for number theory, is not self-contained, that it necessarily includes a statement that is not provable and whose negation is also not provable. As a corollary (his second theorem), Gödel showed that the said system's consistency cannot be proved within the system. In other words, the grand dream to build a self-contained, bottom-up construction of all of mathematics, a goal nurtured by some of the greatest mathematicians of all time, was uncompromisingly shot down. One could, of course, supplement the faulty logical system with extra axioms to prove its consistency, and mathematicians have done so for some systems. But incompleteness had done its damage. After Gödel, the aura of perfection and beauty that defined thousands of years of Platonic realism in its many variations was lost. The dam may not have burst, but the cracks were visible for all to see.

Gödel took aim at the monumental three-volume *Principia Mathematica* by Bertrand Russell and Alfred North Whitehead, produced between 1910 and 1913, whereby the authors attempted to ground all of mathematics in pure logic. The effort was the epitome of perfect rationality. Their goal was to show that symbolic manipulation, once supplied with a set of rules, could capture all of mathematical thinking. Gödel switched symbols for numbers, showing that symbolic patterns in the *Principia* could, in fact, be represented as numerical patterns, or number crunching. Given that Russell and Whitehead's work was self-referential (closed on itself like the fabled Ouroboros), Gödel showed that the whole project echoed issues that had been raised by old logic paradoxes, mainly the Liar's Paradox: "This statement is false."

A bit of reflection shows that this kind of paradox locks reason into an endless loop: the statement can't be true, since, if it is, it asserts that it is false. It can't be false either since, if it is, it asserts the truth. Gödel showed that he could write a formula within the premises of the *Principia* that contradicted itself: "This formula is unprovable by the rules of the *Principia Mathematica.*"[1] How awful for Russell and Whitehead's program, which so valiantly attempted to rid mathematics of such logical vicious circles. As Hofstadter remarked, "In this shockingly bold manner, Gödel stormed the fortress of *Principia Mathematica* and brought it tumbling down in ruins."[2]

Mathematics carries in its very roots the seeds of its own limitations, a hard blow to the pride of many who had believed otherwise, that there was an absolute realm of mathematical truth accessible to the human mind.[3] Curiously, as Rebecca Goldstein argued in *Incompleteness*, her clear and engaging essay on Gödel's work, the general perception of the theorems runs right against what Gödel himself believed—namely, that there is a Platonist realm of mathematical truth that mathematics allows us to probe. Goldstein adds that the same had happened to Einstein, whose belief in a physical reality independent of the human mind was unshaken even after the quantum revolution (see Part 2 of this book), and whose theory of relativity is often perceived as a move against this realist perspective, introducing the less precise human factor into a quantitative description of the world.[4] For Einstein, it was Nature that existed "out there"; for Gödel, it was mathematical purity that existed "out there." For both, the contradiction between realism and idealism, and the limitations on knowledge that idealism imposed, were unacceptable. Our minds should not dictate what the world "out there" is like.

In spite of their revolutionary contributions to their fields, both men spent the last decades of their lives in a sad intellectual exile, talking mostly to one another in their long daily walks around the campus of the Institute for Advanced Study in Princeton. Perhaps, as Goldstein speculates, it was this intellectual exile, so similar for both men, that united them in a strange friendship that lasted until Einstein's death in 1955.

Five years after the publication of Gödel's work, Alan Turing in England introduced the notion of what is now known as a Turing machine, a device able to manipulate symbols on a strip of tape following a set of rules. A Turing machine is, in essence, an idealized computer equipped with a program and an unlimited amount of storage space for computations. In practice, for a finite amount of time and sufficient memory, most computers perform as Turing machines. The device and the tape are what we call the "hardware," the machinelike part of the device, while the set of rules is the program or algorithm. Turing showed that any Turing machine suffers from the "halting problem," its inability to ascertain whether an arbitrary program halts or runs forever. Of course, some programs are easily seen to halt, such as the line of code "print 'Island of Knowledge.'" It will print the statement and be done with it. Others, however, are not: "while (true) continue," where "(true)" can be a statement or list of statements identified as true, such as a number plus itself = twice the number. This program will keep adding number after number and never stop as long as there is energy to run the machine. For more complex programs the decision to halt or not is much more problematic.

The importance of the halting problem, and its connection with Gödel's incompleteness theorems, is that it is an example of an "undecidable" problem, somewhat like the Liar's Paradox. Turing showed that it is impossible to construct a single algorithm that leads to a correct yes-or-no answer to whether the program stops. This is really the crux of the matter, since it means that there will always be propositions whose truth or falsity cannot be decided in a finite number of steps. Inasmuch as mathematics stems from an axiomatic structure following a symbolically implemented set of rules, Gödel and Turing answered the great mathematician David

Hilbert's famous three questions of 1928 in the negative: mathematics as a formal structure is *not* complete, it is *not* consistent, and it is *not* decidable. In other words, the mechanization of human mathematical thought is a mere fantasy.

If the conclusions may be disappointing for those who nurture Platonic dreams of mathematical perfection, for others they are nothing short of wondrous, as they point to the tremendous power of human creativity. The crack in the dam of mathematical perfection exposes the innards of human frailty, ennobling our attempts to construct an ever-growing Island of Knowledge. Gödel and Turing have brought into the open the complex nature of mathematical truth and, by inference, of truth in general. We can't always answer our questions by following a closed set of rules, since some questions are undecidable. In the language we have developed here, the truth or falsity of certain propositions is unknowable. As a consequence—at least within our current logical framework—we can't conceive a system of knowledge constructed with the human brain that is formally complete. Some of our intellectual resources are unruly, springing from novel principles of demonstration and discovery that don't fit the rigid constraints of logic. For *Star Trek* fans, this means we can't ever be like Spock and his fellow Vulcans. How refreshing that we are not slaves to a formal intellectual process! It is this very limitation, and the unexpected intellectual spaces it brings forth in the ongoing struggle for understanding, that makes the pursuit of knowledge all the more unpredictable and thus exciting. Incompleteness leads to creative freedom.

CHAPTER 31

SINISTER DREAMS OF TRANSHUMAN MACHINES: OR, THE WORLD AS INFORMATION

(Wherein we examine whether the world is information,
the nature of consciousness, and whether
reality is a simulation)

The limitations of mathematics as a closed and complete formal system affects another essential area of knowledge: the relationship between machines and human intelligence, a profound and still perplexing scientific question. Can machines think as we do, ultimately to become creative, innovative entities, as opposed to unthinking followers of instructions on a code? Can the human mind, in all of its complexity, be modeled, its essence captured so as to be implementable in machines?

This question can fill (and has filled) many books, and we will not do it justice here. Instead, I'd like to explore, even if partially, how it may inform our discussion on the limits of knowledge and its impact on our humanity.

In the concluding paragraphs of *Gödel's Proof*, Ernest Nagel and James Newman argued that one of the implications of the incompleteness theorems was that computing machines, at least as understood then (the book was first published in 1958), would not be able to emulate the human mind: "There is no immediate prospect of replacing the human mind by robots."[1] They noted that irrespective of their storage capacity and processing speed, machines necessarily follow a linear, step-by-step logic based on a fixed axiomatic method (the program and its syntax), which, as Gödel had shown, was unable to solve innumerable problems in elementary number theory that human brains could. During the past half century, the advent of cellular automata, machine learning, parallel coding, and other techniques have shortened the distance between electronic and human brain processing and thus have changed the situation considerably. Still, the fact remains that "there is no immediate prospect of replacing the human mind by robots."

Machines have been able to beat humans in tasks that apparently ask for intelligent reasoning. For example, IBM's supercomputer Deep Blue defeated world chess champion Garry Kasparov in a rematch in 1997, and its Watson machine beat former *Jeopardy!* winners Brad Rutter and Ken Jennings in 2011. Even if these feats are impressive to some and worrisome to others, there is no deep reasoning or creative thinking going on in these computer victories, at least from the machines: only great programming, tremendously fast processing, and access to prodigiously large data banks. (In the case of Watson, two hundred million pages of content, including the whole of Wikipedia.) If anything, these machines and their victories are triumphs of human ingenuity, not of silicon-based brilliance.

There are different levels of machine intelligence, and there is no question that enormous progress has been made toward emulating certain aspects of human brain functioning. But "strong" artificial intelligence (AI), meaning legitimate machine thinking, remains a long-term goal. One of the key reasons is that we don't quite know what intelligence is or how the human brain (and, to a smaller extent, the brains of other high-functioning animals) is able to sustain it. If intelligence depends solely on the details of the brain's architecture, including the myriad synaptic connections between neurons, then it is reasonable to expect that achieving strong AI is a matter of time: eventually, machines will be able to emulate, and presumably surpass, the human brain. This is the fundamental conjecture that de facto started strong AI research in a 1956 conference at Dartmouth College: "Every aspect of learning or any other feature of intelligence can be so precisely described that a machine can be made to simulate it."[2] If, on the other hand, intelligence and consciousness rely on something else, such as some as yet unknown organizational principle or principles, simple-minded reverse engineering will not be enough to construct thinking machines.

The practical feasibility of thinking machines, then, depends on how the brain functions and the nature of mind. The problem, and the challenge, is that there is widespread disagreement among the experts in these matters. The strong AI proposal, known as "computationalism," assumes that the brain is essentially decodable, that everything comes down to how neurons communicate to one another and build operational clusters: there is no grand mystery of mind, just current ignorance of what the mind's operational principles are. Optimists, such as inventor Ray Kurzweil, robot designer Hans Moravec, and cyberneticist Kevin Warwick, are convinced that in the not too far future computers will be able to simulate the human brain. In 1965, Intel cofounder Gordon Moore obtained an empirical relation now known as Moore's Law: the number of transistors in integrated circuits doubles approximately every two years. Kurzweil has adapted it to modern microprocessor technology to extrapolate that by 2029 desktop computers will have as much processing power as a human brain. More alarmingly, he also speculates that by 2045 artificial intelligence will be able to self-improve at an unthinkable rate, a point in history science fiction writer Vernor Vinge has dubbed "the singularity."

When I was a postdoctoral fellow at Fermilab, Marvin Minsky, one of the founding fathers of strong AI and a signatory of the 1956 Dartmouth Proposal, came to give a colloquium. After presenting his arguments as to why machines would soon be thinking (this was 1986), I asked him if, in that case, they would also develop mental pathologies such as psychosis and bipolar illness. His answer, to my amazement and incredulity, was a categorical "Yes!" Half-jokingly, I then asked if there would be machine therapists. His answer, again, was a categorical "Yes!" I guess these therapists would be specialized debuggers, trained in both programming and machine psychology.

One could argue, contra Minsky, that if we could reverse engineer the human brain to the point of being able to properly simulate it, we would understand quite precisely the chemical, genetic, and structural origin of such mental pathologies and should deal with them directly, creating perfectly healthy simulated minds. In fact, such medical applications are one of the main goals of replicating a brain inside computers. One would have a silicone-based laboratory to test all sorts of treatments and drugs without the need of human subjects. Of course, all this would depend on whether we would still be here by then.

Such sinister dreams of transhuman machines are the stuff of myths, at least for now. For one thing, Moore's Law is not a law of Nature but simply reflects the fast-paced development of processing technologies, another tribute to human ingenuity. We should expect that it will break down at some point, given physical limitations in computing power and miniaturization. However, if the myth were to turn real, we would have much to fear from these codes-that-write-better-codes digital entities. To what sort of moral principles would such unknown intelligences adhere? Would humans as a species become obsolete and thus expendable? Kurzweil and others believe so and see this as a good thing. In fact, as Kurzweil expressed in his *The Singularity Is Near*, he can't wait to become a machine-human hybrid.[3] Others (presumably most in the medical and dental profession, athletes, bodybuilders, and the like) would not be so enthusiastic about letting go of our carbon carcasses. Pressing the issue a bit further, can we even understand a human brain without a human body? This separation may be unattainable, body and brain so intertwined with one another as to make it meaningless to consider them separately. After all, a big chunk of the human brain (the same for other animals) is dedicated to regulating the body and the sensory apparatus. What would a brain be like without the grounding activity of making sure the body runs? Could a brain or intelligence exist only to process higher cognitive functions—a brain in a jar? And would a brain in a jar have any empathy with or understanding of physical beings? Let us hold this thought for a moment.

Although processing speed and access to huge data banks can do wonders to simulate certain aspects of our brains, such attributes are far from being enough to create the totality of mental experiences we call "mind." We can program a machine to recognize the style of different painters and attach certain aesthetical values to each, and then have the machine emit an opinion on a new creation by one artist or another. We can even program the machine to produce a painting that will follow the stylistic lines of a given painter or to compose in the style of Mozart or Bach. We can train the computer to simulate a reaction that we call "emotional" when digitally fed a new painting (a machine doesn't really "see") or a symphony (it doesn't "hear" either). But such reactions are never genuine; they were already there, implanted in the program. The challenge, the open question, is why individual responses to a painting or a piece of music are so different. What personalizes our feelings, our responses to sensorial stimuli that have emotional content? Why are you you?

The concept essential to any argument related to how the mind works amounts to one word: "information." All of physical reality is essentially information, encoded in varying degrees of complexity in the way atoms are coupled together to form the various material structures that compose the world and ourselves. In principle, the brain is no exception. If the computationalists are right, there is a clear reductionist pathway to the mind based on a methodic decoding of its informational content: the brain has so many neurons, networked to one another in this or that way, with a long list of chemicals flowing through its synapses, and so on. Once this informational content is obtained, it can be implemented into the proper hardwiring, and an artificial mind can be built from it, just as a house is built bottom up, from its foundation and walls, to include electrical wiring, plumbing, roof, and, lastly, decor. The fundamental assumption of the strong AI community is that once the human brain is "properly" simulated, the mind will naturally and spontaneously emerge. There is, however, absolutely no empirical evidence for this assumption. In fact, a quick analysis indicates that this belief is somewhat naïve, even cryptoreligious, given what we currently know of how the brain works and the nature of consciousness.

Modern supercomputers are capable of an amazing number of operations per second ("ops" for short). The current record holder (as of July 2013) is Cray's Titan, with 17.59 thousand trillion ops, or 17.59 petaflops. (The prefix "peta" means a one followed by fifteen zeros, or one thousand trillion, represented mathematically as 10^{15}.)[4] The computer has over half a million cores (compare to your laptop boasting a dual-core processor), shared between the usual CPUs and fast graphic processing cards (GPUs) popular in video game platforms. It occupies more than four thousand square feet of space and uses roughly the energy of nine thousand homes. (Believe it or not, this makes the Cray Titan a highly energy-efficient machine compared to its competitors.) Its processing power amounts to roughly three million computations by every human on the planet each second.

There is widespread expectation that supercomputers will soon reach the staggering exaflops mark, or one million trillion ops. (The prefix "exa" means a one followed by eighteen zeros, mathematically 10^{18}.) Optimists such as neuroscientist Henry Markram place this landmark transition as soon as 2018. Markram was recently awarded €1 billion from the European Union to lead the Human

Brain Project, a joint effort of a dozen European organizations to build a full-fledged simulation of the human brain. The project combines cutting-edge neuroscience and computer technology to attempt to implement the myriad details of our brain's architecture and connectivity in a massive computer code. This means incorporating the details of each and every cell type (no two neurons are alike), including its morphology, connectivity, three-dimensional structure, synaptic communication—down to the level of how neurotransmitter molecules travel across ion channels—into the clustering and larger-scale neuronal organization in the different brain regions. Estimates place the computational cost of such a simulation at the exaflop mark: if Markram and the computationalists are successful, exaflop machines will be able to mimic the human brain. There are two essential assumptions here: first, that in the brain, hardware creates software; second, that we will have detailed enough knowledge of all the relevant physiological variables in the brain to feed them into a simulation.

The first assumption seems to be quite natural; after all, what else would there be within our brains apart from the hardware of neurons and their synaptic connections? To imagine that something else is there is to revert to a Cartesian dualism, that a soul-like entity is within us. To do so brings up all sorts of difficulties with modern science, most obviously the question of the soul's immateriality: If the soul is not material, how can it interact with the material? Indeed, if it does interact, it must exchange energy with matter. Such an energy exchange process would make the soul or parts of it material, forcing it to have a detectable signature.

Very few modern scientists and philosophers support the view that the brain has an immaterial component. Nevertheless, there is strong dissent among scientists and philosophers whether we humans are able to understand our own consciousness. Clearly, Markram and other neuroscientists believe that understanding the mind as the seat of consciousness is simply a problem of bottom-up complexity, that reverse engineering the brain is possible and

will lead to an eventual understanding of the mind. A more nuanced view is espoused by Thomas Nagel, Colin McGinn, Noam
Chomsky, Roger Penrose, and, to a lesser extent, Steven Pinker and
others, dubbed collectively as the "New Mysterians." Their view, in
particular as McGinn has put it forward, is that we are "cognitively
closed" to understanding the nature of consciousness: the same way
that a mouse will never recite poetry because of the architecture
and functionality of its brain, human brains have their own cognitive limitations, one of them being understanding consciousness.

The notion is not new. In *Language and Problems of Knowledge*,
Noam Chomsky points out how the limited cognitive powers of all
organisms make for diverse functional abilities: "A Martian scientist, with a mind different from ours might regard this problem [of
free will] as trivial, and wonder why humans never seem to hit on
the obvious way of solving it. This observer might also be amazed
at the ability of every human child to acquire language, something
that seems to him incomprehensible, requiring divine interven-
tion."[5] Nagel explored similar issues in his famous essay "What Is
It Like to be a Bat?," arguing that humans are incapable of experi-
encing how a bat perceives reality through echolocation.[6] In other
words, borrowing from Kant's terminology, what is phenomenon to
one sort of brain is noumenon to others: certain things are beyond
our categories of understanding, the intellectual tools that serve us
well in the study of phenomena.

Echoing Chomsky and Nagel, McGinn's "transcendental naturalism" doesn't rule out that more advanced brains will understand
consciousness: the problem is not unanswerable in principle, it is just
unanswerable by us at this point in our evolutionary development.

Going back 145 years, here is how eminent Victorian physicist
John Tyndall put it in his 1868 presidential address to the Physical
Section of the British Association for the Advancement of Science:

The passage from the physics of the brain to the corresponding
facts of consciousness is unthinkable. Granted that a definite

thought, and a definite molecular action in the brain occur simultaneously, we do not possess the intellectual organ, nor apparently any rudiment of the organ, which would enable us to pass by a process of reasoning from the one phenomenon to the other. They appear together but we do not know why. Were our minds and senses so expanded, strengthened and illuminated as to enable us to see and feel the very molecules of the brain, were we capable of following all their motions, all their groupings, all their electric discharges, if such there be, and were we intimately acquainted with the corresponding states of thought and feeling, we should be as far as ever from the solution of the problem. How are these physical processes connected with the facts of consciousness? The chasm between the two classes of phenomena would still remain intellectually impassable. . . . Let the consciousness of love, for example, be associated with a right-handed spiral motion of the molecules of the brain, and the consciousness of hate with a left-handed spiral motion. We should then know, when we love, that the motion is in one direction, and, when we hate, that the motion is in the other; but the "Why?" would remain as unanswerable as before.[7]

Clearly, Tyndall would have opposed Markram's project. The essence of the Mysterian argument is that some problems are simply too difficult for us to solve, given our intellectual abilities. Such mysteries speak directly to the limits of knowledge and, in some cases, to the existence of unanswerable questions, regions of the unknowable out there in the ocean of the unknown.

The Mysterian critique of the computationalist view goes something like this: there is a clear confusion between the *physiology* of thinking—the phenomenal choreography of neurons flashing and neurotransmitters flowing—and the *substance* of the thinking process, what the thinking is *about*. As McGinn recently put it, "When you look at a painting or read a poem your brain no doubt undergoes electrochemical activity; but the painting or poem is not *in* your brain. . . . The artwork is the *object* of the mental act of apprehending it; it is not the mental act in which it is apprehended."[8] McGinn and other Mysterians place the burden of proof on the computationalists: Can *they* show that the experiential mind is traceable to the collective flow of neuronal computations in the brain? Can *they* explain how subjective experience arises from neural computations? McGinn thinks the mission is impossible and blames it on the way our perception of the brain constrains the way we can make sense of its functioning: consciousness is not an observable quality that can be pinned down; it's not *in* this or that part of the brain or in this or that particular neuronal process. Its trade name is elusiveness.

The problem of consciousness is so hard that it can't even be formulated properly. Australian philosopher David Chalmers, presently at New York University, has famously called this "the hard problem of consciousness," as he distinguished it from the "easier" ones, such as the differences between being asleep and being awake, or how sensorial information is processed and integrated by a cognitive system.[9] The quotation marks serve as a reminder that these "easy" problems are extremely difficult as well; the difference is that they are amenable to the usual methods of the cognitive neurosciences, while the hard problem isn't. Although most scientists and philosophers agree that understanding consciousness is difficult (unless they dismiss the question altogether), some would claim that although it should be clear that there are cognitive limitations to what our brains can fathom, we cannot be sure that indeed we are cognitively incapable of understanding how the mind works.[10]

Either way, these are philosophical arguments, and even if some are quite convincing and the debate essential, they don't amount to a proof. We don't have what physicists like to call a "no-go" theorem. In the absence of one, it is extremely difficult to determine with final certainty which questions are categorically unanswerable. "Never" is a hard word to use in science. If we're trying to make some headway into the issue, it is perhaps useful to shift gears a little and attempt to relate the hard problem of consciousness to the nature of reality and how we understand it.

The question of consciousness is deeply related to the notion of reality. Even if other animals have consciousness and interact with physical reality in deliberate ways, presumably we are the only terrestrial species that have self-awareness *and* a high enough level of cognitive complexity to contemplate the nature of consciousness, even if we may end up perplexed by it. Self-awareness coupled to a high level of cognitive complexity leads to the unique human faculty of self-consciousness, the ability to reflect on our own existence.

We exist in a world that we believe is real. By "real" I mean a world that is not a fabrication of our minds, that has an existence independent of how we perceive it. This belief is built from the integration of sensory stimuli that come from the "outside"—that is, from the world "out there"—to the world within. I am thus taking a stance against the radical idealist who believes that everything is rooted in the mind and that no external reality exists without it. Reality "out there" exists, even if its nature depends on how we perceive it "in here." The physical pain of kicking a stone (the stone is really there, even if each individual feels pain in a unique way), or the billions of years of cosmic history without minds (intelligence, human and otherwise, takes a while to evolve) to think about it, are, to me, convincing evidence that the world exists independently of us.

People, of course, may disagree as to the exact details of the world out there, and hallucinations can greatly distort those details. But there is *something* out there, a reality wherein each one of us exists and that our brains capture in part through our sensory apparatus. When I see a blue ball moving, different parts of my brain act together to create the certainty that a blue (color) ball (shape) is moving (movement). This kind of construction is entirely classical, in the sense that quantum effects are entirely (or mostly) negligible: what we usually call our "reality" is a decohered reality. Efforts to use quantum mechanics and its weirdness to explain consciousness or even lower brain functions must deal with the fact that the brain is a warm and wet environment, and thus a hard place to maintain any kind of quantum entanglement. As MIT's cosmologist Max Tegmark and others have argued, decoherence timescales are extremely fast in the brain, shorter than the relevant timescales for neuronal flickering.[11] Even if there are other ways in which quantum mechanics may possibly play a role in the functioning of the brain—for example, in the opening and closing of synaptic gates or in optimizing energy transport across synapses—these will *probably* not be of fundamental relevance in explaining how consciousness emerges through neural activity. I say "probably" because it is best to keep an open mind, given how much we don't know about the inner workings of the brain.

If we thus consider for now that consciousness is a classical entity and that our conception of reality comes from our sensory interactions with the world and our collected memories, can we be sure that reality is real? This is a key point to us. And the shocking answer is a categorical no!

We have seen that what we call "physical reality" depends very much on how we look at the world and on what we know about the world. For the Greeks, and all the way until after Copernicus late in the sixteenth century, the cosmos was Earth-centered and finite, with the sphere of the stars demarking its spatial limits. "Reality" was thus defined, and this definition had profound theological implications, shaping the way people lived their lives. With the discoveries of Edwin Hubble in the 1920s, the cosmos was found to be expanding, and reality took a different bend: the cosmos became a dynamic entity, endowed with its own history. As had happened many times before, reality was redefined, and we are still grappling with what it means to live in a Universe that had a beginning and will have an end.[12]

If, up to the early twentieth century, the doctrines of religious faith dictated to a large extent how people lived their lives, and thus had an enormous emotional and existential impact on society as a whole, in our times it is science that increasingly plays this role. At the most fundamental level, our scientific discoveries define what we call reality. And here comes the twist. We have explored how science, in its effort to explain the workings of Nature, has intrinsic limitations in its precision and in the formulation of natural laws. As these necessarily change in time due to the methodic advancement of human inquiry, the arena we call reality is always shifting. New concepts of space and time, of what a field is and how it shapes the ways matter interacts; the very concept of what matter *is*; even the uniqueness of our Universe: all of these foundational stones of what philosophers call our ontology, the conceptual entities by which we describe reality, are always transitional. *The very nature of scientific inquiry, always ongoing and always under revision, necessarily implies the notion of a changing understanding of reality.* As a consequence, we can't ever state what reality is. The best that we can do is to state what we know of the nature of reality today. Those who cling to the notion that one day we will arrive at the very fundamental essence of reality are victims of what I call The Fallacy of Final Answers, which have plagued human knowledge since Thales first asked what the world was made of.

There is another, more perverse reason why we should distrust the notion that we can comprehend the fundamental essence of reality. Beyond our own limitations in trying to make sense of it, we may be victims of a prank of truly cosmic proportions: reality, or what we think reality is, could simply be a massive simulation, powerful enough to completely fool us into believing it is real. Given that we perceive reality with limited sensorial input and precision, could a virtual copy of reality be simulated so as to be identical to the original? In other words, could we be living in a virtual reality and what we think is reality is just a giant computer code?

First, we must agree that the starting point is perceived reality, that is, the reality we infer as our brains integrate the sum total of our sensory input. As such, simulations don't need to go into extremely fine detail beyond what we can notice.[13] Scientists have a term for this, "coarse-graining," which basically means that we can dispense with details that aren't going to be noticed or that are unimportant—we can smooth over invisible kinks and bumps. This would be "reality" as the Chained Ones in Plato's cave perceive it, which, incidentally, is how we perceive it. Of course, the simulation would have to take into account our observational tools and the ways in which they augment our perception of reality with finer detail. To keep on fooling us as we improve our tools, the simulations would have to keep on increasing their resolution.

In 2003, philosopher Nick Bostrom published an essay in which he considered whether we were already living in a simulation.[14] Assuming that posthuman civilizations (presumably of the kinds that would emerge beyond Kurzweil's *Singularity* and that would have the cognitive openness to figure out the mind-body problem) would have computational resources greatly exceeding our own, Bostrom concluded that the question of whether we are already living in a simulation can have three possible answers, two negatives and one positive: (1) Human civilization goes extinct before reaching the "posthuman" phase (doomsday scenario). (2) Posthuman civilizations have no interest in running simulations of their ancestors (psychological scenario). (3) We *are* living in such a simulation. Given our present ignorance, Bostrom rates the three answers as equally likely. If the answer is 3, and we are living in a simulation, the mystery of consciousness is akin to the puppet hopelessly wondering who moves the strings that move him. How horribly powerless we would be in this case! Bostrom's assumption, and that of the many science fiction books and movies exploring this theme (most famously the Wachowskis' movie *The Matrix*, although in the same year, 1999, there was also Josef Rusnak's *The Thirteenth Floor*, addressing similar themes) is that with enough computer power, reality can be simulated in a way to completely fool us into thinking that our lives are real, that our perceived reality is real.[15]

There is an even worse scenario. In Bostrom's case we would still exist as flesh-and-blood creatures fooled by an amazingly powerful simulation: stimuli would still be coming into our heads. However, a truly sophisticated simulation need not be restricted to outside stimuli; interior stimuli would also be part of it, including all of our thoughts and dreams. Such a simulation, I am supposing, would be able to mimic even the experience of countless mental states; it would simulate consciousness. Could it be that the reason we are so perplexed by consciousness is because it *is* a simulation and thus seemingly impenetrable or even magical to us?

In this scenario we wouldn't exist as flesh-and-blood creatures (chained to a virtually simulated "cave"), being only characters in a simulation. Sounds preposterous, I know, but consider the popular video game *The Sims*. The name of the game already indicates that it is a simulation, in this case of characters engaged in everyday activities such as relationships, going to school, having babies, and so forth. The player controls the characters and what they do. At the current level of sophistication the characters, of course, are unaware of being part of a game. In fact, they have no awareness whatsoever. But now imagine a future version of the game where characters do have awareness; they are self-conscious. They believe that they exist and that their surroundings are real. The "players" can tweak the level of awareness from primitive to highly sophisticated. With enough sophistication, the characters can be fooled into thinking that they are real, that their lives are real. This being the case, even their free will would be due to a series of code lines, an illusion of a perceived freedom that, in fact, doesn't exist. This game would simulate our existence, including our individual consciousness. And we wouldn't have a clue.

There is something deeply disturbing about this argument, that we may not be masters of our lives but simply animated puppets in the "hands" of other masters. We can't really call the situation sad since the characters (us!) would not know of their slavery: they would consider themselves as real and as free as we do. Could our enormously detailed sensorial perception of the world, our passions and thoughts, our triumphs and defeats, the sweetness and the bitterness of the whole of human experience, even our experience of consciousness, be but an artificial construction of advanced intelligences? This is a question of feasibility, of what it would take to create such kinds of simulations.

There is also a question of motivation. Would such intelligences really find it fun to play such games, or necessary to research their ancestors' simple lifestyles through simulations? Would a posthuman intelligence *need* to have fun? Or is Bostrom's second answer— posthuman civilizations have no interest in running simulations of their ancestors—the one that prevails, so that such kinds of simulations are the marks of primitive intelligences such as ours? (Note that Bostrom's answers also work for the case when we are not just *in* a simulation but *are* simulations.)

Given the above, I favor Bostrom's second answer as the most probable. As a bonus, it is nicely consistent with McGinn's transcendent naturalism, since our posthuman successors may have cracked the mystery of consciousness (they need not be as cognitively closed as we are).

For those who believe we are living in a simulation, here is an interesting twist: there seems to be a circularity in Bostrom's argument. Who is to say that our masters are not themselves part of a simulation? That even more advanced intelligences tricked them into thinking that they are the master puppeteers, while they are, in fact, puppets like the rest of us? A dream within a dream within a dream . . . echoing Edgar Allan Poe's famous lines, "Is all that we see or seem/But a dream within a dream?"

If we are going this far, we would need to consider that the whole universe might be a giant simulation. At this point we would embark in speculative astrotheology, given that intelligences capable of simulating whole universes would be indistinguishable from gods. Or are there limits to how far the simulation game can be played, the question of feasibility mentioned above? Many scientists, most prominently MIT's Seth Lloyd, have equated the universe to a giant computer, stating that, in essence, every physical process—from the collision between two electrons right after the Big Bang to the rotation of the Milky Way to the thoughts I am having right now—is a computation between bits of matter, a transfer of information according to the laws of quantum mechanics: "Every detail that we see around us, every vein on a leaf, every whorl on a fingerprint, every star in the sky, can be traced back to some bit that quantum mechanics created. Quantum bits program the universe."[16]

Lloyd's proposal is that the rich complexity we see in Nature stems from the alliance of two parts: a computer—in this case, the Universe as it busily processes information—and randomness—in this case, quantum decoherence, which provides the random bits that generate short program chunks, that is, little parts of computer programs. Contrary to monkeys typing on typewriters—a process that generates only gibberish—random program strings of varying lengths can, according to the mathematical theory of algorithmic information, generate "all the order and complexity that we see."[17]

If the Universe is a computer, can a computer generate the Universe? Before we consider whether the Universe we live in is a simulation, we need to consider how hard generating the Universe would be. There are stringent physical limits to how much energy and information you can encode and shuffle around in matter. These limits apply to anyone who builds computing machines, human or alien. Every computation involves the manipulation of information through a medium, be it made of matter (as in conventional computer chips) or radiation (photons). In most cases, this means flipping the spin of a magnetic material from "up" to "down" or some similar process. Using quantum physics, we can estimate the number of elementary logical operations an optimal (perfect) device can perform using a fixed amount of energy. If all of the energy of the device (meaning its entire mass using the $E = mc^2$ formula) could be used for the computation, a two-pound optimal laptop could perform a maximum of about 10^{50}

operations in one second (ops).[18] Compare this to our forthcoming supercomputers performing at exaflops, or 10^{18} ops! But speed of computation is not everything; energy and temperature also limit the amount of information that the device can store and process, its memory capabilities. In general, a collection of N two-state systems has 2^N accessible states and can register N bits of information. (The two states here may refer to the "up" and "down" spins of magnetic particles.) This bound comes from the system's entropy, which limits its amount of storage.

In essence, the entropy of a system counts the number of its accessible states, that is, the number of states that can be used for storing information. The higher the entropy of the system, the more information it can store: a checkerboard with twelve squares on a side can "store" many more configurations than one with six squares on a side. For the model two-pound perfect laptop, this amounts to about 10^{31} bits of available memory space. These results can be scaled out to the whole known Universe, assuming that all of it can be used for computation and that it has been since the Big Bang. In 2002, Lloyd estimated that the Universe could have performed 10^{120} ops on 10^{90} bits (or even more, 10^{120} bits if we include gravitational interactions).[19] So if our puppet masters are indeed running a simulation as big as the known Universe, this is what their computer would need to be capable of. They might be able to save a lot of ops and bits by coarse-graining, but the numbers would remain truly staggering. If they tried to save too much, the quality of their simulation would suffer, and we the simulated might catch a glitch in their program by uncovering something weird with our "reality," a tear in the seams of the world. For example, Silas Beane, Zohereh Davoudi, and Martin Savage have speculated that if our simulators use a square lattice to mimic the Universe, a sort of three-dimensional checkerboard with a certain "lattice size" (the size of a square in a normal checkerboard) then their simulation would be limited by how small they manage to make the lattice size. Very high-energy events, which naturally probe very small distances, could test the resolution of the simulation, possibly even unveiling its fake nature.[20]

Combining these arguments with the incompleteness proofs of Gödel and Turing and the unavoidable limitations of self-referential logical systems they expose, we see that even idealized computers can only model to reasonable accuracy a physical system of which they are not a part. There is no perfect, seamless simulation. Furthermore, and most importantly, they would fail if they attempted to model a part of the world that includes themselves.[21]

The point is that even highly sophisticated futuristic simulators will encounter physical limits constraining what they can and can't do. First, their knowledge of reality will be necessarily limited. Second, their ability to simulate their version of reality, that is, to reproduce their knowledge in a machine, will have to conform to the limits of energy resources, processing speeds, and memory storage. A race capable of using the whole Universe to compute for their purpose would appear to be indistinguishable from what we now call God. It is thus refreshing to know that even such beings will have physical limits to what they can and can't do: they won't be gods after all. More to the point, as we further develop our knowledge of the physical universe and the power of our computational resources, we should expect that we will eventually be able to perform computations and other actions that at present would appear to be magical. As Arthur C. Clarke famously said, "Any sufficiently advanced technology is indistinguishable from magic."[22] That the best we can ever do will always amount to a coarse-grained reconstruction of physical reality means that we will never be godlike, just as our purported simulators cannot be. The laws of Nature and the limits of knowledge will make sure that we remain human and fallible.

CHAPTER 32

AWE AND MEANING

(Wherein we reflect on the urge to know and why it matters)

The grand narrative that is the scientific enterprise must be celebrated as one of the greatest achievements of the human intellect, a true testimonial to our collective ability to jointly create knowledge. Science is a response to our urge to understand, to make sense of the world we live in and of our place in it. It addresses the same age-old questions that have haunted and inspired humanity throughout the ages, questions of origins and endings, of place and meaning. We need to know who we are; we need to know where we are and how we got here. Science speaks directly to our humanity, to our quest for light, ever more light.

If reason is the tool we use in science, it is not its motivation. We don't attempt to understand the world as an end in itself. Our search defines us, imprinted as it is with what makes us human: the passion and the drama, the challenges, the experience of elation and defeat, the perennial itch to move on, the disturbing but teasing sense that we know so little—that wonders await, hidden from view, tantalizingly close, yet so mysterious.

We probe Nature the best way we can, with our tools and intuition, with our models and approximations, with our imaginative descriptions, metaphors, and imagery. The view of science I presented here is a view of open-ended pursuit, not of envisioned ends.

As we learn more about the world, confronting theories with data, probing deeper and further, we realize that the answers we gather are steps that go mostly forward but sometimes back: the Island of Knowledge grows and sometimes shrinks, as we learn something new about the Universe or take something back. We see ever more clearly but never clearly enough.

It is too simplistic a hope to aspire to complete knowledge. Science needs to fail in order to move forward. We may crave certainty but must embrace uncertainty in order to grow. We are surrounded by horizons, by incompleteness. All we see are shadows on cave walls. Yet it is also too simplistic to consider such limits as insurmountable obstacles. Limits are triggers: they teach us something about the world and how we perceive it; they teach us something about ourselves while taunting us to keep edging forward, in search of answers. We push the limits and keep on pushing so that we can better know who we are. The same ongoing growth process that we see in science—forward, backward, but always charging ahead—we should see in each of us, in our individual pursuits. The day we become too afraid to step into the unknown is the day we stop growing.

Science is more than just knowledge of the natural world. It is a view of life, a way of living, a collective aspiration to grow as a species in a world filled with mystery, fear, and wonder. Science is the blanket we pull over our feet at night, the light we turn on in the dark, the beacon reminding us of what we are capable of doing when we work together in pursuit of a common goal. That science may be used for good or ill only reflects the stubborn precariousness of human nature and its propensity to create and destroy.

As we probe Nature and master so many of its facets, it is good to remember that the shores of ignorance grow as the Island of Knowledge grows: the ocean of the unknown feeds on our successes. It is also good to remember that science only covers part of the Island, that there are many ways of knowing that can and should feed on one another. While the physical and social sciences surely can illuminate many aspects of knowledge, they shouldn't carry the burden of having all the answers. How small a view of the human spirit to cloister all that we can achieve in one corner of knowledge! We are multidimensional creatures and search for answers in many, complementary ways. Each serves a purpose, and we need them all. Sharing a glass of wine with a loved one is more than just the chemistry of its molecular composition, the physics of its liquid consistency and the light reflections on its surface, or the biology of its fermentation and our sensorial response to it. To all that we must add the experience of its ruby color and of its taste, the pleasure of the company, the twinkle in the eyes across the table, the quickening of the heart, the emotion of sharing the moment. Even if many of these reactions have a cognitive and neuronal basis, it would be a mistake to reduce them all to a measurable data set. It all sums up; it all becomes part of what it means to be alive, to search for answers, for companionship, for understanding, for love.

Not all questions have answers. To hope that science will answer all questions is to want to shrink the human spirit, clip its wings, rob its multifaceted existence. And given this book's explorations of the limits of scientific knowledge, it is also a deeply misguided hope. It is one thing to search for answers to questions of origins and endings, of meaning and purpose within the scientific framework—*that* we must do, always. It is what I have done for most of my life as a scientist. It is another to actually believe that the search has an end, that the ocean of the unknown is bounded and that science alone can chart its expanse. How arrogant it is to claim that we can know it all, that we will be able to pry open all of Nature's secrets one after another like nested Russian dolls until we decipher the very last one! To accept the incompleteness of knowledge is not a defeat of the human intellect; it doesn't mean we are throwing in the towel, surrendering. It means that we are placing science within the human realm, fallible even if powerful, incomplete even if the best tool we have for describing the world. Science is not a reflection of a God-given truth, made of discoveries plucked from a perfect Platonic realm; science is a reflection of our very human disquietude, of our longing for order and control, of our awe and fear at the immensity of the cosmos.

We don't know what lies beyond our horizon; we don't know how to think of the initial state of the Universe or how to obtain a deterministic description of the quantum world. These unknowns are not simply a reflection of our current state of ignorance or of our limited tools of exploration. They express Nature's very essence, bounded by the speed of light, by time's arrow, by an irrevocable randomness. There is an essential difference between "we don't know" and "we can't know." Answers to any of these unknowns, even if found, would be of limited range. Unless we can travel faster than light, we can't directly probe beyond our cosmic horizon. Any scientific answer to the initial state of the Universe depends heavily on the conceptual scaffolding of the scientific framework—fields, conservation laws, uncertainties, and the nature of space, time, and gravity—and quantum nonlocality defies any hope of having a deterministic explanation of the world of the very small. More generally, *any* scientific explanation is necessarily limited.

I understand that it is hard for some to accept that this limitation doesn't take away from the beauty and explanatory power of science. This resistance, I believe, is rooted in an antiquated way of thinking, in which science is seen as the conqueror of all mystery, a view that confuses the fictional goal of acquiring absolute knowledge with the drive to keep on searching. Quite the contrary, I argue: to see science for what it is makes it more beautiful and powerful, not less. It aligns science with the rest of the human creative output—impressive, multifaceted, and imperfect as we are.

Even around us, within our immediate territory, we only see a fraction of what's there. Recall that we are immersed in dark matter and dark energy, that the stuff we are made of amounts to just 5 percent of what fills the cosmos. At this moment in history, unknown dark materials surround us. Even if our tools keep on improving, as they surely will, even if we finally unlock the mystery of dark matter and dark energy, as I am sure we will, there is only so much information we can gather. The unexpected is doubtlessly lurking right under our noses, with its potential to foster a deep shift in our current worldview.

The map of what we call reality is an ever-shifting mosaic of ideas.

Take note once again, reader, that mine is not a view of defeat, of giving up on the quest. Quite the contrary, the quest *must* go on, always. The quest is what makes us matter: to search for more answers, knowing that the significant ones will often generate surprising new questions. When science is seen within the historical perspective that I presented here, it is quite simple not only to come to terms with the incompleteness of knowledge but to embrace it as the ever-reaching-forward symbol of what being human is all about. For it is our prodigious appetite for wonder that feeds our prodigious ability to probe the new.

Like our ancestors, we bow to the vastness of the undertaking, to the beauty that hides in the unknown, beckoning us. It is awe that fuels us and that has fueled us from the start. Awe is the bridge between our past and our present, taking us forward into the future as we keep on searching. Let us not trivialize this argument to a matter of "we can know it all" versus "we can't know it all." Let us, instead, embrace the awe in our hearts and minds, embrace the impulse to learn, to discover, to shine a bit more light ahead, to push the boundaries of the Island of Knowledge forward, sideways, inwards, push them anywhere—as long as we keep on pushing toward greater understanding. Let us rage, rage against the dying of the light, never conceding to go gentle into that good night. To keep on shining: this is what matters. This is what we are here for.[1]

Acknowledgments

The idea for writing this book came during the conference *Laws of Nature: Their Nature and Knowability*, which took place in May 2010 at the Perimeter Institute for Theoretical Physics in Canada. The organizers, Steve Weinstein, David Wolpert, and Chris Fuchs, were kind enough to invite me and to let me choose what to talk about. I am endlessly grateful to them for prompting me to elaborate on my ideas on the limits of science and the nature of knowledge.

It was while thinking about what to say to a very distinguished group of scientists and philosophers that the image of an "Island of Knowledge" came to me, together with its immediate consequences: that we are surrounded by an ocean of unknowns, and that, as the Island grows so do the shores of our ignorance, and thus our ability to ask questions we couldn't have anticipated. Another issue that the Island image prompted was whether there are unknowables—that is, whether there are questions outside the reach of scientific thinking. The positive response from my colleagues and the many conversations that ensued after my presentation were kindling to my imagination. The result, after four years of labor, is the present book.

I would like to thank my many colleagues who lent their time and wisdom to my prodding questions about their views on knowledge. But first and foremost, my heartfelt thanks go to Adam Frank, David Kaiser, and Nicole Yunger-Halpern for having read the complete book manuscript and for their invaluable comments and criticisms. We all know what a precious commodity time is nowadays, when so many enticing lights are flashing nonstop all around us.

I also thank my agent, Michael Carlisle, for believing in this project from the very start and my editor at Basic Books, T. J. Kelleher, for making it become a reality.

Lastly, I thank my five children, Andrew, Eric, Tali, Lucian, and Gabriel, for teaching me to shine a spotlight where life really matters and to look at the world with renewed wonderment every day. And my wife, Kari, for her sustaining love, support, and understanding.

Notes

PROLOGUE

1. The statement "the smallest bits of stuff that make up all that exists in the world" needs careful unpacking, and I will do so in detail in Part 2. We must ask whether scientists can ever know for sure that they have found the "smallest bits of stuff." As we shall see, this question is directly related to the limits of knowledge.

2. The analogy must be taken with a grain of salt, since colliding oranges at mundane speeds are quite different from colliding particles of matter at speeds close to the speed of light. The creation of new kinds of particles is a direct product of the conversion of energy of motion into mass. Unless you accelerate an orange sufficiently close to the speed of light, all you will see flying out of the collision will be guts, juice, and cracked seeds. Physicists are fond of saying that colliding particles close to the speed of light would be like colliding two tennis balls and getting out Boeing 747 jets.

3. "Elementary" here refers to indivisible, as in "not composed of anything smaller." (See note 1.) The quotes indicate that we must take the assertion of a particle as being "elementary" with suspicion. A more appropriate statement would be: given our *current* understanding of the properties of matter, this or that particle can be considered "elementary," or structureless. The emphasis on "current" is essential.

4. Science, of course, is only one way to "know more than we can see." Art complements it as an attempt to alleviate the blindness of the emotional world within, to bridge between the elusive realm of feelings and the more palpable through words, images, and sounds.

5. Bernard le Bovier De Fontenelle, *Conversations on the Plurality of Worlds* (Berkeley: University of California Press, 1990), 1.

6. As I was doing a final bibliography check before sending my manuscript to the editors, I came across a notion very similar to my own Island of Knowledge, by famous Austrian physicist Victor Weisskopf: "Our knowledge is an island in the infinite ocean of the unknown, and

the larger this island grows, the more extended are its boundaries toward the unknown," Victor Weisskopf, *Knowledge and Wonder: The Natural World as Man Knows It* (Garden City, NY: Doubleday, 1962), quoted in Louise B. Young, ed., *The Mystery of Matter* (New York: Oxford University Press, 1965), 95. Weisskopf doesn't elaborate further on this idea, while I will do so here. Science journalist John Horgan, in his controversial *The End of Science: Facing the Limits of Knowledge in the Twilight of the Scientific Age* (New York: Broadway Books, 1996), 83, attributes a similar statement to American physicist John Archibald Wheeler: "As the island of our knowledge grows, so do the shores of our ignorance."

Another image of remarkable similarity to my own, without the notion of an island, came to my attention when this work was half-completed. Sir William Cecil Dampier, in his *A History of Science and Its Relations with Philosophy and Religion*, 4th ed. (Cambridge: Cambridge University Press, 1961), wrote: "There seems no limit to research, for as been truly said, the more the sphere of knowledge grows, the larger becomes the surface of contact with the unknown" (500). I thank my *13.7* blog reader "Mark I" for bringing it to my attention, even if unaware of my project. The notion of an island or sphere of knowledge is clearly compelling. Indeed, the metaphor appears, perhaps for the first time, in German philosopher Friedrich Nietzsche's *The Birth of Tragedy*: "For the periphery of the circle of science has an infinite number of points; and while there is no telling how this circle could ever be surveyed completely, noble and gifted men nevertheless reach, e'er half their time, and inevitably, such boundary points on the periphery from which one gazes into what defies illumination" (*Basic Writings of Nietzsche*, trans. Walter Kaufmann [New York: Modern Library, 2000], 97).

PART I

CHAPTER 1: THE WILL TO BELIEVE

1. Later on, I will carefully distinguish between this intangible kind of unknowable and what I call "scientific unknowables," which happen to be a crucial part of our understanding of Nature.

2. Mircea Eliade, *Images and Symbols: Studies in Religious Symbolism* (New York: Sheed & Ward, 1961), 59.

3. It behooves the scientist, in no easy feat of professional integrity, to let go of this faith when observations no longer support it. Letting go is hard to do.

4. Isaac Newton, *The Principia: Mathematical Principles of Natural Philosophy*, trans. I. Bernard Cohen and Anne Whitman (Berkeley: University of California Press, 1999), 796. Indeed, in the "third rule for the study of natural philosophy" Newton proposes that "the qualities of bodies [that cannot be increased and diminished] and that belong to all bodies on which experiments can be made should be taken as qualities of all bodies universally" (795).

CHAPTER 3: TO BE, OR TO BECOME?
THAT IS THE QUESTION

1. Aëtius quoted in Daniel W. Graham, ed., *Texts of Early Greek Philosophy: The Complete Fragments and Selected Testimonies of the Major Presocratic* (Cambridge: Cambridge University Press, 2010), Part 1, 29.

2. Graham, *Texts of Early Greek Philosophy*, Part 1, 35.

3. Isaiah Berlin, "Logical Translation," in *Concepts and Categories: Philosophical Essays*, ed. Henry Hardy (New York: Viking, 1979), 76.

4. Graham, *Texts of Early Greek Philosophy*, Part 1, 55.

5. See, for example, Carlo Rovelli's biography of Anaximander, *The First Scientist: Anaximander and His Legacy* (Yardley, PA: Westholme, 2011).

6. Graham, *Texts of Early Greek Philosophy*, Part 1, 47.

7. Graham, *Texts of Early Greek Philosophy*, Part 1, 57.

8. A right at least of male individuals. With the exception of the Pythagoreans, who alone included women on an equal standing.

9. Upon reading lines like these we see why Stephen Greenblatt, in his masterful *The Swerve: How the World Became Modern*, attributed to Lucretius and his poem a key role in the launching of the modern age.

10. G. S. Kirk, J. E. Raven, and M. Schofield, *The Presocratic Philosophers: A Critical History with a Selection of Texts*, 2nd ed. (Cambridge: Cambridge University Press, 1983), 343.

11. Nicolaus Copernicus, *On the Revolutions of the Heavenly Spheres*, trans. Edward Rosen (Baltimore: Johns Hopkins University Press, 1992), 4–5.

CHAPTER 4: LESSONS FROM PLATO'S DREAM

1. Plato, *The Dialogues: The Republic, Book VII*, trans. Benjamin Jowett, Great Books of the Western World, vol. 7, ed. Mortimer J. Adler, 2nd ed. (Chicago: Encyclopaedia Britannica, 1993), 389, line 517.

2. Lucretius, *The Nature of Things, Book II*, trans. A. E. Stallings (1060; rept., London: Penguin, 2003), 67–68.

3. That every hypothesis must fail sooner or later is a consequence of the way science evolves: through a constant revision of how it models and describes Nature. What an electron was in the late nineteenth century is quite different from what it was in the 1940s or what it is now. We will have the opportunity to revisit this key point as this book advances.

4. We can see here the origin of the notion of God as the Cosmic Watchmaker, prevalent among eighteenth-century Deists such as Benjamin Franklin.

5. Simplicius of Cilicia, *On Aristotle's "On the Heavens 2.1–9,"* trans. Ian Mueller (Ithaca, NY: Cornell University Press, 2004), 74 (line 422,20).

6. Moses Maimonides (1135–1204), "The Reality of Epicycles and Eccentrics Denied," trans. Shlomo Pines, in *A Source Book in Medieval Science*, ed. Edward Grant (Cambridge, MA: Harvard University Press, 1974), 517–520.

7. Graham, *Texts of Early Greek Philosophy*, 83.

8. The fact that we still use the word "meteor-ology" to describe the weather speaks to the tremendous influence of Aristotle's ideas in Western culture. Clouds and thunderstorms have little to do with meteors!

9. Martin Luther, *Table Talk*, Luther's Works, vol. 54, trans. and ed. Theodor G. Tappert (Philadelphia: Fortress, 1967), 358–359.

CHAPTER 5: THE TRANSFORMATIVE POWER OF A NEW OBSERVATIONAL TOOL

1. J. L. E. Dreyer, *Tycho Brahe* (Edinburgh, 1890), 86f.

2. This kind of across-the-sky motion, called "proper motion," was first noticed by Edmund Halley, of eponymous comet fame, in 1718. Stars may also move toward and away from us, that is, in radial motion, detected using the Doppler effect, the slight change in the wavelength of light waves (the distance between two successive crests or troughs) as their source moves toward (wavelengths shorten) or away (wavelengths increase) from the observer.

3. Recall that from a terrestrial perspective the Sun moves along the sky, completing one revolution in a year. As it does so, it runs by the twelve constellations of the Zodiac, the ones appearing in horoscopes. Because the Earth spins around itself with a tilt of 23.5 degrees like a falling top, the Sun's circular path is tilted by that same amount, going above and below the celestial equator. Hence the name "right ascension." The vernal and autumnal equinoxes are the points where the Sun's path

crosses the celestial equator, the points where these two imaginary circles join at zero right ascension.

4. Angles follow the same sexagesimal pattern as hours, minutes, and seconds. So as one hour can be divided into sixty minutes, an angle of 1 degree can be divided into 60 arc minutes (one arc minute being then 1/60th of 1 degree), and an angle of 1 arc minute can be divided into 60 arc seconds (1 arc second being then 1/3,600th of one degree).

5. To this avid fly fisherman nothing gives more joy than to know I'm in such august company.

6. With a simple exercise you can see how parallax works. Stretch your arm out and close your left eye. Now look at your thumb and at an object farther away, say a painting on the wall. Notice their relative positions. Now close your right eye and notice again their positions: your thumb moved but the painting hardly did. In Tycho's case, the two eyes were the locations of the two astronomers (Denmark and Prague), the Moon was the thumb, and the comet the faraway painting.

7. As I wrote much about Kepler's life in some of my other works, we will take leave of his many vicissitudes here, focusing instead on his science.

8. To keep things in perspective, I should note that "sharp" here is a bit of an exaggeration. To visualize the departure of Mars's orbit from a perfect circle, we could draw the orbit and the circle on a fifty-foot billboard. Mars's orbit would bulge out by just one inch.

9. Harriot's moon drawings can be seen online at "Thomas Harriot's Moon Drawings," *The Galileo Project*, 1995, http://galileo.rice.edu/sci/harriot_moon.html. For a biography see John W. Shirley, *Thomas Harriot: A Biography* (Oxford: Clarendon, 1983).

10. Galileo should have been aware that Tycho's model was also compatible and, in fact, predicted, the phases of Venus. However, he chose to overlook this, as he did, regrettably, Kepler's elliptical orbits.

11. From Kepler's manuscript on the Supernova of 1604, *De Stella Nova*, quoted in Alexandre Koyre, *From the Closed World to the Infinite Universe* (Baltimore: Johns Hopkins University Press, 1957), 61.

CHAPTER 6: CRACKING OPEN THE DOME OF HEAVEN

1. Even though there is some debate as to whether Galileo actually performed this experiment, as you enter the tower, there is a plaque celebrating the occasion; further, Galileo's pupil and first biographer, Viviani, claimed that the incident indeed took place. In any case, I performed the experiment at the famous tower for a Brazilian TV series on the history of science. Repeatability is essential to the scientific trade.

2. Here is the link to the YouTube video: http://www.youtube.com /watch?v=KDp1tiUsZw8. How amazed would Galileo be if he knew that his experiment was repeated on the surface of the Moon, and in less than four hundred years since his own?

3. Jonathan Hughes, *The Rise of Alchemy in Fourteenth-Century England: Plantagenet Kings and the Search for the Philosopher's Stone* (London: Continuum, 2012), 24.

4. Newton, *Mathematical Principles*, 941.

5. Blaise Pascal, *Pensées*, trans. A. J. Krailsheimer (New York: Penguin, 1995), nos. 205 and 206.

CHAPTER 7: SCIENCE AS NATURE'S GRAND NARRATIVE

1. Isaac Newton, *Four Letters to Richard Bentley*, in *Newton: Texts, Backgrounds, Commentaries*, ed. I. Bernard Cohen and Richard S. Westfall (New York: Norton, 1995), 330–339.

2. Newton, *Mathematical Principles*, 943.

3. Note that my argument has nothing to do with traditional schisms in philosophy such as relativism, or postmodernism versus positivism, or any claim that science is essentially subjective, or, at the other extreme, that it is the only way to the truth. Even if scientific concepts often come from the subjective musings of individuals or groups of individuals within a certain cultural context, in their practice scientists aim at universals, that is, at results verifiable and repeatable by those willing and equipped to do so. The essential point here is that the scientific description of reality is an ongoing, self-correcting narrative process, with a central commitment to efficacy. We could call my take on the philosophy of science a *natural constructivism*, a point I will elaborate on later.

CHAPTER 8: THE PLASTICITY OF SPACE

1. But take note: light does travel with different speeds in different mediums, say, vacuum versus air or water. It tends to slow down in a denser medium. For example, light's speed inside a diamond is only about 41 percent of its speed in vacuum.

CHAPTER 9: THE RESTLESS UNIVERSE

1. In Einstein's words, "If we are concerned with the [metrical] structure only on a large scale, we may represent matter to ourselves as being uniformly distributed over enormous spaces, so that its density of

distribution is a variable function which varies extremely slowly." Albert Einstein, *Cosmological Considerations on the General Theory of Relativity* [1917], in *The Principle of Relativity: A Collection of Original Papers on the Special and the General Theories of Relativity*, trans. W. Perrett and G. B. Jeffery (New York: Dover, 1952).

2. The Hooker hundred-inch telescope held the distinction of being the world's largest from 1917 to 1948. It was named after John D. Hooker, a Los Angeles businessman who provided the funds for the giant hundred-inch mirror.

3. Without the steam locomotive and the higher speeds it could attain, the demonstration of Doppler's idea would have been quite difficult. The context of discovery depends in essential ways on the available technology.

4. Robert Schulmann, A. J. Kox, Michel Janssen, and József Illy, eds., *The Collected Papers of Albert Einstein*, vol. 8, *The Berlin Years: Correspondence, 1914–1918* (Princeton, NJ: Princeton University Press, 1998), Document 321.

5. In my book *The Dancing Universe* I explore in great detail the history of twentieth-century cosmology. Here I am focusing mostly on the ideas we will need to make sense of things later on.

6. Incidentally, this is roughly when the Sun will be entering its red giant phase, whereupon it will inflate and engulf Mercury and Venus and graze Earth's orbit. Even if galactic collisions are less dramatic than they sound (stars are very far apart, and the chance of stellar collisions is remote), the end of the Sun does signal the end of Earth as a life-bearing planet.

CHAPTER 10: THERE IS NO NOW

1. Considering that light travels 983,571,056 feet in 1 second, to cover 1 foot it will take 1/983,571,056 second, or 1.0167×10^{-9} second. Light travels 1 foot in 1 billionth of a second in empty space—a good relation to remember. (Even though air is not empty space, the change in the speed of light is very tiny.)

2. In the cognitive neurosciences there is great interest in understanding when the brain perceives a sensory input and, for example, how visual and auditory signals are perceived as simultaneous when they are not. (Say, when a ping-pong ball bouncing on a table is seen and heard.) J. V. Stone et al. reported that the notion of simultaneity breaks down for visual and audio signals at different times for different people—that is, you and I perceive audio-visual simultaneity differently—although there is larger agreement if light precedes sound by fifty-two milliseconds (J.

V. Stone et al., "When Is Now? Perception of Simultaneity," *Proceedings of the Royal Society of London [B]* 268 [2001]: 31–38). Furthermore, it appears that we may react to a visual stimulus before being aware of it. In other words, unless the visual stimulus is fairly complex, awareness doesn't always guide action. See, e.g., J. Jolij, H. S. Scholte, S. van Gaal, T. L. Hodgson, and V. A. Lamme, "Act Quickly, Decide Later: Long-Latency Visual Processing Underlies Perceptual Decisions but Not Reflexive Behavior," *Journal of Cognitive Neuroscience* 23, no. 12 (2011): 3734–3745. However, we must also consider that our current understanding of "awareness" is not as clear as it should be to detect it.

CHAPTER 11: COSMIC BLINDNESS

1. More precisely, what I am calling "light" here is not only visible light but all kids of electromagnetic radiation, of which visible light is but a tiny fraction. The electromagnetic spectrum runs from radio waves, with the longest wavelength (lowest frequency and hence lowest energy), to microwaves to infrared to visible to ultraviolet to x-rays to gamma rays, with the shortest wavelength and hence highest energy.

2. To avoid clutter, unless specified I will use "light" generically to represent any type of electromagnetic radiation.

3. The light atomic nuclei present at this time were synthesized between one-hundredth of a second and three minutes after the Big Bang, during an epoch called "nucleosynthesis." They include a few isotopes of hydrogen (deuterium and tritium, with one proton and one and two neutrons in the nucleus, respectively), of helium (helium-3 and helium-4, with two protons and one and two neutrons, respectively), and lithium-7 (with three protons and four neutrons). Larger atomic nuclei were only synthesized hundreds of millions of years later, from the exploding remains of dying stars.

4. Given that electrons and protons were never combined into hydrogen atoms before this time, I find this choice of the word "recombination" somewhat confusing.

5. Numbers are taken from the Planck satellite team's best-fit analysis. See, e.g., http://arXiv.org/abs/1303.5082.

6. That galaxies may actually be carried away faster than the speed of light does not violate Einstein's theory of relativity, even if it appears to do so. The speed of light imposes a limit on how fast information or particles of matter may travel, but it doesn't limit the speed at which space itself may stretch.

7. George Gordon (Lord) Byron, "Darkness," in *The Works of Lord Byron: A New, Revised, and Enlarged Edition with Illustrations*, ed. Ernest Hartley Coleridge, vol. 4 (London: John Murray, 1901), 42.

CHAPTER 12: SPLITTING INFINITIES

1. Many books pro-superstrings (e.g. by Brian Greene, Michio Kaku, Leonard Susskind) and against (Lee Smolin, Peter Woit) have been written and are listed in the bibliography. The subject remains fascinating, even if current data is going strongly against some of its assumptions, such as the existence of supersymmetry. In any case, our work here is to explore the nature of physical reality within what is known in science, and not what is still merely speculative, even if compelling.

CHAPTER 13: ROLLING DOWNHILL

1. An often-used and pertinent example is a falling elevator: the faster it accelerates down, the lighter you'd feel. If the elevator free falls, you'd feel weightless.

2. In Newton's theory, only the density of the gas contributes to the pull of gravity. This fact makes for an enormous difference in how the two theories model the evolution of the cosmos.

3. To avoid unnecessary complication, I will use the terms "metastable" and "phase transition" rather loosely here.

4. You may reasonably object that times of trillionths of a second are way too small to be relevant. For us, perhaps. But for elementary particles, such timescales are very noticeable. For example, in one-trillionth of a second a photon can travel one-third of a millimeter, a huge distance in particle physics equivalent to crossing about five million hydrogen atoms.

5. Indeed, "false vacuum" is somewhat of a misnomer, since the notion of false vacuum only applies if the matter is trapped in a higher energy state and needs an energy kick to get down to its lower energy state. Think of a basketball stuck to the hoop that will only come down to its lowest energy state—at the ground—if someone hits it hard enough. The picture the reader should keep in mind is that of a ball that can roll up or down a hill, not always with an obstacle stopping it from rolling down. Hence the use of "displaced energy" instead of "false vacuum."

CHAPTER 14: COUNTING UNIVERSES

1. This is the OED's second possible definition, the first being related to William James's original use of the term in his 1895 article "Is Life Worth Living?": "Visible nature is all plasticity and indifference, a multiverse, as one might call it, and not a universe" (*International Journal of Ethics* 6 [October 1895]: 10). James's definition of multiverse is what we call Universe, and we will not consider it any further.

2. The reader should not confuse the current accelerated expansion, fueled presumably by dark energy, with the primordial accelerated expansion of inflationary cosmology. The early acceleration gave way to a slower expansion for about five billion years, when the current phase took off.

3. Mary-Jane Rubenstein's *Worlds Without End: The Many Lives of the Multiverse* (New York: Columbia University Press, 2013) offers a thorough and accessible survey of the many kinds of multiverse proposed in the history of cosmological thought.

CHAPTER 15: INTERLUDE: A PROMENADE ALONG THE STRING LANDSCAPE

1. The *M* originally stood for "membrane," a generalization of possible surfaces that could also be fundamental and include the one-dimensional strings, but now the *M* stands variously for mystery, magic, or mother, according to Witten himself.

2. Lisa Randall, *Warped Passages: Unraveling the Mysteries of the Universe's Hidden Dimensions* (New York: Harper Perennial, 2005).

3. Several books that address the Anthropic Principle are listed in the bibliography, such as John Barrow and Frank Tippler's *The Anthropic Cosmological Principle*, Paul Davies's *Cosmic Jackpot*, and Sir Martin Rees's *Before the Beginning*. In my own *A Tear at the Edge of Creation*, I spend considerable time explaining the shortcomings of the Anthropic Principle as a predictive tool in the physical sciences. There are at least two versions of the principle, the strong and the weak. The strong version rings of cosmic teleology, as it states that the cosmos is such that we *should* be here, and will not be considered any further.

4. I took this example from my good friend Alex Vilenkin's book *Many Worlds in One*. However, I use it in the precise opposite way that Vilenkin does, as I point out the shortcomings and not the virtues of anthropic reasoning.

CHAPTER 16: CAN WE TEST THE
MULTIVERSE HYPOTHESIS?

1. George Ellis, "Does the Multiverse Really Exist?" *Scientific American* (August 2011).

2. I am thus proposing what philosophers of science may like to call a "natural constructivism," a doctrine in which scientific theories are not discoveries of timeless truths about the cosmos but ongoing and ever-changing human constructions based on a careful balance between what can be observed through our measuring tools and our ability to devise mathematical models to describe what we see. Our best theories are those that fit the data, even if we cannot be certain that they are unique. In any case, we should be certain that they are not final.

3. Indeed, a few days after I wrote these lines, an article with this title appeared in *Discover* magazine, written by Steve Nadis.

4. The double peak in the polarization pattern was first suggested in a paper by Matthew Kleban, Thomas S. Levi, and Kris Sigurdson, "Observing the Multiverse with Cosmic Wakes," September 15, 2011, http://xxx.lanl.gov/pdf/1109.3473.pdf. I can't resist mentioning that Tom Levi was my undergraduate advisee at Dartmouth and that his first paper was our joint collaboration.

5. More precisely, and keeping the image of a bubble bath, imagine sprinkling black pepper over the froth: the grounds would be distributed around "voids," the insides of the bubbles. In the Universe, galaxies are distributed in a similar fashion, around voids with little or no matter inside.

PART II
CHAPTER 17: EVERYTHING FLOATS IN NOTHINGNESS

1. I am considering only Western ideas of unification. There are, of course, many unifying principles in Eastern religious and philosophical traditions, from Buddhism and Hinduism to Taoism and, to a larger or smaller extent, these may have influenced the thoughts of some of the Presocratics.

2. Western philosophers were not alone in proposing atoms. In India, Buddhists, Jains, and Hindu philosophers espoused different versions of atomism. The Jains, in particular, and earlier than the Greeks, proposed a thoroughly materialistic version of atomism whereby each atom had one kind of taste, smell, and color, and two states—subtle (able to fit into the

smallest spaces) and gross (larger). Atoms even had a property similar to opposite electric charges, which led to an attractive property that allowed them to bind. It is not clear whether such ideas influenced Western atomists, although Diogenes Laertius, a historian living in the third century CE, relates that Democritus went to India and met with Gymnosophists (extreme ascetics who despise food and clothing as detrimental to pure thoughts).

3. Democritus, Fragment 32c, quoted in Graham, *Texts of Early Greek Philosophy*, 597.

4. Democritus, Fragment 40, quoted in Graham, *Texts of Early Greek Philosophy*, 597.

5. Epicurus, *Letter to Herodotus* 1.85−87, http://www.college.columbia .edu/core/sites/core/files/text/Letter%20to%20Herodotus_0.pdf.

6. Epicurus, *Letter to Pythocles* 1.32−34, http://www.epicurus.net/en /pythocles.html.

CHAPTER 18: ADMIRABLE FORCE AND EFFICACY OF ART AND NATURE

1. Note that this is not the same as immortality: death could still come through violence or accident. One should contrast this alchemical belief with modern searches for gene therapy, organ cloning, and other means of extending human life combining biology with digital technology. Both use the cutting-edge science of the time to address the universal human need to find a solution to death. Mary Shelley's *Frankenstein* also fits squarely within this tradition, as it explored the newly discovered taming of electricity and its power to move muscles as a possible way to beat death.

2. Roger Bacon, *The Mirror of Alchimy: Composed by the Thrice-Famous and Learned Fryer, Roger Bachon*, ed. Stanton J. Linden (New York: Garland, 1992), 4.

3. Jared Diamond, *Guns, Germs, and Steel: The Fates of Human Societies* (New York: Norton, 1997).

4. The hardness of bronze as compared to copper is due to its lattice structure. While copper is organized in a regular lattice, the atoms of tin added to make bronze break this regularity and act as a sort of blockade, restricting the movement of the copper atoms and making it harder to break the lattice structure down.

5. There is some controversy as to whether Jabir ibn Hayyan was the discoverer of *aqua regia* and many of the cited acids. Zygmunt S. Derewenda claims he was in his article on the history of tartaric acid ("On Wine, Chirality and Crystallography," *Acta Crystallographica* A64 [2008]: 246−258). Other historians of science claim that the discoverer

was probably an European alchemist under the pseudonym Pseudo-Gerber, often associated with the thirteenth-century Italian monk Paul of Taranto, specially after detailed research by William R. Newman. Whatever the resolution, Jabir's enduring fame through the Middle Ages is obvious from the choice of the pseudonym. His works, real or apocryphal, became the staple for European alchemists for over five centuries. Readers interested in an elucidating discussion on the disputes concerning the nature of alchemy, including its role as a spiritual and occult practice, should consult Lawrence M. Principe and William R. Newman, "Some Problems with the Historiography of Alchemy," in *Secrets of Nature: Astrology and Alchemy in Early Modern Europe*, ed. William R. Newman and Anthony Grafton (Cambridge, MA: MIT Press, 2001).

6. Quoted in Eric John Holmyard, *Makers of Chemistry* (Oxford: Clarendon, 1931), 60.

7. B. J. Dobbs, "Newton's Commentary on the Emerald Tablet of Hermes Trismegistus," in *Hermeticism and the Renaissance*, ed. Ingrid Merkel and Allen G. Debus (Washington, DC: Folger Shakespeare Library, 1988).

8. Platinum is less reactive than gold; however, being rather rare, it didn't have as large a role as it could. Some 80 percent of its current world production is from South Africa.

9. Bacon, *The Mirror of Alchimy*, 4.

10. William R. Newman, "The Alchemical Sources of Robert Boyle's Corpuscular Theory," *Annals of Science* 53 (1996): 571.

11. Adapted from Jane Bosveld, "Isaac Newton, World's Most Famous Alchemist," *Discover* (July–August 2010), http://discovermagazine.com /2010/jul-aug/05-isaac-newton-worlds-most-famous-alchemist.

12. Newton, *The Principia*, 938.

13. Ibid., 382–83.

14. Isaac Newton, *Opticks* (London: William Innys, 1730), Query 8.

15. Ibid., Query 31. These quotes are from the end of a long speculative text, in which Newton expounds his amazing knowledge of chemistry, accumulated from years of alchemical experiments.

16. Ibid., Query 30.

CHAPTER 19: THE ELUSIVE NATURE OF HEAT

1. The three relations (actually, any pair of them) yield the combined gas law $PV = kT$, where P is pressure, V is volume, T is temperature, and k is an arbitrary constant.

2. In formulas, $T \approx v^2$, and $P \approx n\ v^2$, where v^2 is the average velocity squared and $n = N/V$ is the "number density" of molecules, the ratio

of their number N to the volume V. Even before Waterston, in 1820 the English physicist John Herapath had proposed that the momentum (velocity times mass) of a particle in a gas is a measure of the temperature of the gas. Although the correct relation is to the square of the temperature, Herapath's ideas were published in the *Annals of Philosophy*, after being rejected by the Royal Society. The atomistic hypothesis was clearly floating about, although few were taken by it.

3. Benjamin (Count Rumford) Thompson, "An Experimental Enquiry Concerning the Source of the Heat Which Is Excited by Friction," *Philosophical Transactions of the Royal Society* (1798): 102.

CHAPTER 20: MYSTERIOUS LIGHT

1. The theoretical explanation for the blueness of the sky is known as "Rayleigh scattering," after the British physicist Lord Rayleigh, who showed that the intensity of the scattered light was inversely proportional to the fourth power of its wavelength ($I \sim 1/\lambda^4$). Since blue light has shorter wavelength than others in the visible spectrum, it is scattered the most; that is, it is spread about the most.

2. This complicates things a bit, since water and sound waves are longitudinal—that is, they oscillate in the same direction as they propagate. The transversal nature of light waves confused scientists for quite a while.

3. Thomas Young, "An Account of Some Cases of the Production of Colors Not Hitherto Described" (1802), reprinted in *The Wave Theory of Light: Memoirs by Huygens, Young and Fresnel*, ed. Henry Crew (New York: American Book, 1900), 63–64.

4. There were many other experiments that also returned negative results. I mention the Michelson-Morley one since it is the best known. Results could be at first order in the ratio of v/c, where v is the velocity of motion with respect to the aether, or to second order, v^2/c^2. First-order results could be explained away if the aether as a whole drifted. But second-order results, as in the Michelson-Morley experiment, presented a serious challenge to the aether idea.

5. Albert Einstein, *On the Electrodynamics of Moving Bodies*, reprinted in *The Principle of Relativity*, 37.

6. Ibid., 38.

7. Albert Einstein, "On a Heuristic Point of View About the Creation and Conversion of Light," in *The Old Quantum Theory: Selected Readings in Physics*, by D. ter Haar (New York: Pergamon, 1967), 104.

8. Ibid., 92.

CHAPTER 21: LEARNING TO LET GO

1. Albert Einstein, "Does the Inertia of a Body Depend upon Its Energy-Content?" reprinted in *The Principle of Relativity*, 71.

2. Ibid.

3. More technically, to go upstairs you do work against the attractive gravitational field of the Earth. The amount of work you do on the climb equals the amount of potential gravitational energy gained. Going down, you release that extra potential energy. The electron must do work to climb away from the attractive grip of the proton.

4. Specifically, de Broglie associated a wavelength λ to a body of mass *m* and velocity *v* and thus of momentum *p* = *mv* according to the relation λ = *h*/*p*, where h is Planck's constant. The formula can be refined for objects moving with relativistic speeds, that is, speeds close to the speed of light:

$$\lambda = \frac{h}{mv} \sqrt{1 - \frac{v^2}{c^2}}$$

When *v* is much smaller than *c*, the formula reduces approximately to the former expression. Note that as *v* increases, the particle's wavelength shrinks in accordance with relativistic length contraction.

5. Max Born, *The Born-Einstein Letters: Correspondence Between Albert Einstein and Max and Hedwig Born from 1916–1955, with Commentaries by Max Born*, trans. Irene Born (London: Macmillan, 1971), 91.

6. Anton Zeilinger, *Dance of the Photons: From Einstein to Quantum Teleportation* (New York: Farrar, Strauss and Giroux, 2010), 78.

CHAPTER 23: WHAT WAVES IN
THE QUANTUM REALM?

1. Superfluids are a good illustration of macroscopic quantum behavior: quantum cooperative effects at low temperatures may allow a fluid to flow with very little viscosity. "Cooperative effects" means that many atoms behave in tandem so as to amplify the effect to macroscopic scales. Superfluid helium creeps up on the walls of its containing vessel as if some magical force moved it against gravity.

2. Schrödinger to Lorentz, in *Letters on Wave Mechanics: Schrödinger, Planck, Einstein, Lorentz*, trans. Martin J. Klein (New York: Philosophical Library, 1967), 55.

3. For readers who are familiar with complex numbers, the wavefunction $\psi(t,x)$ is actually a complex function. To obtain the probability, a

real quantity, we must compute its absolute square, that is, multiply it by its complex conjugate, $\psi^*(t,x)$. Since the electron can be anywhere in space, we must also make sure that the wavefunction is well behaved, that is, its absolute square drops to zero at spatial infinity, $\psi^*(t,x)\psi(t,x)\rightarrow 0$ as $x\rightarrow\pm\infty$. To be able to make sense as a probability, the wavefunction needs to be normalized, $\int \psi^*(t,x) \psi(t,x)dx = 1$. (The particle needs to be somewhere in space!) We can then define the probability of finding the electron at point x and time t as $P(x,t) = \psi^*(t,x) \psi(t,x)$. The solution to Schrödinger's equation is the wavefunction $\psi(t,x)$. From it we can compute $P(x,t)$.

4. Let's say the electron could be found in one of four positions, x_1, x_2, x_3, and x_4. Before its position is measured, it could be in any of these four with some probability, and its wavefunction reflects this. Detection would mean the electron would be found at one of the four options. Let's say it was found at x_2. After detection, its wavefunction would be $\psi(x_2,t)$. (Of course, it will never be exactly at x_2 since every measuring device has limited accuracy. But it will be in the neighborhood of x_2 within the precision of the device.)

5. This analogy is only meant to be suggestive: a snake is a real object, while a wavefunction isn't. Also, the collapse of the wavefunction appears to be instantaneous, while the coiling of the snake around a single rung takes time.

CHAPTER 24: CAN WE KNOW WHAT IS REAL?

1. Quoted in Max Jammer, *The Philosophy of Quantum Mechanics: The Interpretations of Quantum Mechanics in Historical Perspective* (New York: Wiley, 1974), 151.

2. Unless, like physicist John Wheeler, we believe that we can influence histories backward in time, a possibility we will get back to soon enough. Wheeler goes as far as suggesting that our current existence influenced cosmic history so that the Universe could exist for us to develop in it.

3. Albert Einstein, Boris Podolsky, and Nathan Rosen, "Can Quantum-Mechanical Description of Physical Reality Be Considered Complete?" *Physical Review* 47 (1935): 777–780.

4. Mathematically this means that the order of their product gives the same results, as in $2 \times 4 = 4 \times 2 = 8$. Quantities like this are said to be commuting (or compatible): you can measure them in any order, and the results are the same. Incompatible quantities in quantum mechanics do not commute: their order does affect the end result. This is not as strange as it seems. Even in our reality there are noncommuting quantities—for

example, rotating a book in two different nonparallel directions by large enough angles. The reader can verify that inverting the order of the rotations takes the book to a different final position. (Position the book facing you and imagine three axes going through it. Choose two of them and rotate the book clockwise around one axis and then the other. Note where the book ends. Now return the book to its original position, revert the order of the rotations, and voilà!)

5. For example, if the two equal mass particles were emitted by a decaying particle at rest, we know that their speeds must be the same and in opposite directions. This is due to the conservation of momentum: if the momentum is zero initially (source particle at rest), it will remain zero afterwards (two particles moving in opposite directions). This sort of decay is quite common in particle physics. Or one could use light: photons always travel at the speed of light in empty space.

6. Niels Bohr, "Can Quantum-Mechanical Description of Physical Reality Be Considered Complete?" *Physical Review* 48 (1935): 696−702.

7. David Bohm, *Quantum Theory* (1951; rept., New York: Dover, 1989), 620.

8. Ibid.

9. If we use the Greek letter ψ to represent the cat's overall wavefunction, quantum theory would prescribe it to be written as a superposition of two possible states, ψ_{alive} and ψ_{dead}, corresponding to the cat being alive or dead, respectively. The formula would read $\psi = a\psi_{alive} + b\psi_{dead}$, where a and b are number coefficients that have the added complication of being complex numbers. They are composed using the number i, which is $\sqrt{-1}$. So, $i^2 = -1$. A typical complex number (z) is written in terms of two real numbers (x and y) as: $z = x + iy$. The "absolute value" of a complex number is always positive and defined as follows: $|z|^2 = z z^* = (x + iy)(x - iy) = x^2 + y^2$. The probability of the cat being alive is given by $|a|^2$, the absolute value squared of a; of it being dead, $|b|^2$. Initially, when the cat is put in the box alive, $|a|^2 = 1$: the probability of it being alive is 100 percent. Once the box is closed, the cat goes into a superposition of both states. If, when the box is opened, the cat is still alive, then $|a|^2 = 1$ as before. If the cat is dead, then $|a|^2 = 0$ and $|b|^2 = 1$.

10. Entangled wavefunctions are usually represented as sums of the product of the entangled entities. Suppose a detector is set up to measure the spin of an electron and that the spin can only be either + (up) or − (down). *Before* the measurement, the wavefunction for the electron is $\psi_{el} = \psi_{el}(+) + \psi_{el}(-)$ (neglecting constant numerical factors). For the detector, the wavefunction is $\psi_{detector}$. The detector also has (at least) two states, measuring the electron's spin either up or down. The joint wavefunction

for the electron and the detector is $\psi = \psi_{\text{detector}} [\psi_{\text{el}}(+) + \psi_{\text{el}}(-)]$. They are entangled. The electron, in a sense, is both spin-up and spin-down. Or you could say it has no definite spin. *After* the measurement, when the electron "collapsed" into a definite state, the joint wavefunction is *either* $\psi = \psi_{\text{detector}} \psi_{\text{el}}(-)$ or $\psi = \psi_{\text{detector}} \psi_{\text{el}}(+)$. The key point is that while two entities are entangled, it makes no sense to describe them by their individual wavefunctions. Only the entangled state will do. The act of measurement destroys the entangled state as it pins down the electron's spin.

11. For a delightful and informative history of how the interpretation of quantum mechanics has taken new wings in the late 1960s and 1970s see David Kaiser's *How the Hippies Saved Physics: Science, Counterculture, and the Quantum Revival* (New York: Norton, 2011).

12. Zeilinger's *Dance of the Photons*, a popular book on his experiments and the weirdness of quantum mechanics, is a joy to read.

CHAPTER 25:
WHO IS AFRAID OF QUANTUM GHOSTS?

1. The joint wavefunction for the two entangled photons would be something like: $\psi = \psi^A_v \psi^B_v - \psi^A_h \psi^B_h$, where A refers to Alice's photon and B to Bob's, while v and h refer to the photons being either in a vertical or in a horizontal polarization state. Don't worry about the minus sign.

CHAPTER 26: FOR WHOM THE BELL TOLLS

1. Bohm, *Quantum Theory*, 115.

2. David Bohm, "A Suggested Interpretation of the Quantum Theory in Terms of 'Hidden' Variables: I," *Physical Review* 85, no. 2 (1952): 166.

3. John S. Bell, *Speakable and Unspeakable in Quantum Mechanics* (Cambridge: Cambridge University Press, 1987), 160.

4. Seth Lloyd, *Programming the Universe: A Quantum Computer Scientist Takes on the Cosmos* (New York: Knopf, 2006).

5. What follows is not the setup in Bell's paper but a simpler variation based on the so-called CHSH inequality. CHSH stands for the papers' four authors, J. F. Clauser, M. A. Horne, A. Shimony, and R. A. Holt, "Proposed Experiment to Test Local Hidden-Variable Theories," *Physical Review Letters* 23, no. 15 (1969): 880–884.

6. For example, the experimentalist could run the experiment one thousand times, recording the results in a table like this:

	(L|,R|)	(L|,R/)	(L/,R|)	(L/,R/)
RUN 1	(+,−)	(−,−)	(−,+)	(−,+)
RUN 2	(−,+)	(−,+)	(−,−)	(+,−)
. . .	. . .	. . .	. . .	. . .
RUN 1000	(−,+)	(−,−)	(+,−)	(−,+)

7. For each run, the value of C can be computed. For example, RUN 1 in the table above would give:

$$C_1 = C(\text{RUN } 1) =$$
$$(+-)-(-+)+(--)+(-+)=(-1)-(-1)+(+1)+(-1)=0$$

8. Marissa Giustina et al., "Bell Violation with Entangled Photons, Free of the Fair-Sampling Assumption," *Nature* 497 (May 9, 2013): 227–230.

CHAPTER 27: CONSCIOUSNESS AND THE QUANTUM WORLD

1. In a fitting version for our times, the 2011 movie *Limitless* explores similar effects induced by a super-Ritalin pill.

2. Randi has been a decisive voice against all sorts of dishonesty in the field of psychic claims. In one video he debunks both Uri Geller and the shameless Evangelist healer Peter Popoff: http://www.youtube.com /watch?v=M9w7jHYriFo. In 2009, Uri Geller reversed his claim that he had any kind of special psychic powers, calling himself a "mystifier" and entertainer. As for starting broken watches, research has shown that over 50 percent of watches brought in for repair are not mechanically broken but stopped because dirt or gummed oil impaired the mechanism. A large fraction of these watches start ticking for a brief time when held in closed warm hands and shaken. See, e.g., David Marks and Richard Kammann, "The Nonpsychic Powers of Uri Geller," *Zetetic* 1 (1977): 9–17; James Randi, *The Truth About Uri Geller*, rev. ed. (New York: Prometheus Books, 1982).

3. David Kaiser, *How the Hippies Saved Physics*, provides an engaging account of these Victorian gentlemen and how modern claims of psychic ability and their presumed connection with quantum mechanics have much in common with the Victorians' ideas.

4. Maximilian Schlosshauer, Johannes Kofler, and Anton Zeilinger, "A Snapshot of Foundational Attitudes Toward Quantum Mechanics," *Studies in History and Philosophy of Science Part B: Studies in History and Philosophy of Modern Physics* 44, no. 3 (August 2013): 222–230.

5. Eugene Wigner, "Remarks on the Mind-Body Question," reprinted in *Quantum Theory and Measurement*, ed. John Archibald Wheeler and Wojciech Hubert Zurek (Princeton, NJ: Princeton University Press, 1983), 169.

6. Ibid., 177.

7. Ibid., 173.

8. C. M. Patton and J. A. Wheeler, "Is Physics Legislated by Cosmogony?" in *Quantum Gravity: An Oxford Symposium*, ed. C. J. Isham, R. Penrose, and D. W. Sciama (Oxford: Clarendon, 1985), 538–605.

9. Ibid., 564.

10. V. Jacques et al., "Experimental Realization of Wheeler's Delayed-Choice Gedanken Experiment," *Science* 315 (2007): 966–968. I also mention two recent experimental verifications of Wheeler's delayed choice using entangled photons and thus adding explicit nonlocality to the mix: F. Kaiser et al., "Entanglement-Enabled Delayed-Choice Experiment," *Science* 338 (2012): 637–640, and A. Peruzzo et al., "A Quantum Delayed-Choice Experiment," *Science* 338 (2012): 634–637.

11. J. A. Wheeler, "Law Without Law," in Wheeler and Zurek, eds., *Quantum Theory and Measurement*, 182–213.

12. Ibid., 197.

13. Ibid., 199.

14. David Deutsch, *The Beginning of Infinity: Explanations that Transform the World* (New York: Penguin, 2011), 308.

15. James Hartle, "The Quantum Mechanics of Closed Systems," in *Directions in General Relativity*, vol. 2 (Festschrift for C. W. Misner), ed. B. L. Hu, M. P. Ryan, and C. V. Vishveshwara (Cambridge: Cambridge University Press, 1993). The shorter version, where this quote came from, can be found in http://xxx.lanl.gov/pdf/gr-qc/9210006.pdf.

16. Bell, *Speakable and Unspeakable in Quantum Mechanics*, 171.

CHAPTER 28: BACK TO THE BEGINNING

1. My guess is that it cannot; life needs a level of order and continuity that is impossible within the randomness of the quantum realm. Life may emerge at the transition between the quantum and the classical, and surely many known and yet unknown quantum effects play roles in its functioning. But living systems need the consistency of classical physics to be viable.

2. For experiments, see J. R. Reimers, L. K. McKemmish, R. H. McKenzie, A. E. Mark, and N. S. Hush, "Weak, Strong, and Coherent Regimes of Fröhlich Condensation and Their Applications to Terahertz Medicine and Quantum Consciousness," *Proceedings of the National Academy of Sciences* 106, no. 11 (2009): 4219–4224. For theory, see M. Tegmark, "Importance of Quantum Decoherence in Brain Processes," *Physical Review E* 61, no. 4 (2000): 4194–4206.

PART III
CHAPTER 29: ON THE LAWS OF HUMANS
AND THE LAWS OF NATURE

1. I note that many physicists, starting with Paul Dirac, have proposed that the laws of Nature need not be strictly time-independent but could vary in time. For example, some of the fundamental constants of Nature could be slowly varying over eons, a small and thus very difficult effect to measure. In my early days as a researcher I investigated whether theories with extra spatial dimensions, such as those motivated by superstring theory, could produce time-dependent fundamental "constants"; the answer is yes, although there are very stringent observational limits on their changes—so stringent, in fact, that they behave as constant over cosmological times, that is, times of billions of years. More recently, João Magueijo and collaborators have proposed that the speed of light may have varied in the past, and Lee Smolin has proposed that the laws of Nature may change across the Big Bang singularity. Both ideas still need to be empirically validated. I list their engaging books in the bibliography. To us, time-varying constants or laws of Nature are excellent illustrations of how the limitations of measurement allow for new ideas to flourish in science: given that we can only measure the values of the constants to a certain level of precision, there is always room for change underneath what we can grasp.

2. T. G. Hardy, *A Mathematician's Apology* (1940: rept., Edmonton: University of Alberta Mathematical Sciences Society, 2005), 23, http://www.math.ualberta.ca/mss/misc/A%20Mathematician%27s%20Apology.pdf.

3. Ibid., 41.

4. George Lakoff and Rafael E. Núñez, *Where Mathematics Comes From: How the Embodied Mind Brings Mathematics into Being* (New York: Basic Books, 2000), xvi. In 1998, while Lakoff and Nuñez were working on their book, George Johnson wrote a very informative essay for the *New York Times*, putting together the evidence for math as invention, citing

many different sources, from Lakoff and the mathematician Gregory Chaitin to neuroscientists; see George Johnson, "Useful Invention of Absolute Truth: What Is Math?" *New York Times*, February 10, 1998, http://www.nytimes.com/1998/02/10/science/useful-invention-or-absolute-truth-what-is-math.html.

5. Quote from interview with Gregory Chaitin by Robert Lawrence Kuhn, "Is Mathematics Invented or Discovered?" video, Closertotruth.com, http://www.closertotruth.com/video-profile/Is-Mathematics-Invented-or-Discovered-Gregory-Chaitin-/1433 (accessed August 9, 2013).

6. Michael Atiyah, "Created or Discovered?" video, Web of Stories, http://www.webofstories.com/play/michael.atiyah/88;jsessionid=36092 DC06C8A5D5C2C2E755A2CD70972 (accessed June 25, 2013).

7. Atiyah's position seems to waver, as he admits this is a difficult question without an obvious resolution; he has also proposed the parable of the lonely jellyfish (at least that's what I call it), which gives credence to the "invention" camp:

We all feel that the integers really exist in some abstract sense and the Platonic view is extremely seductive. But can we really defend it? It might seem that counting is really a primordial notion. But let us imagine that intelligence had resided, not in mankind, but in some vast solitary and isolated jellyfish, buried deep in the Pacific Ocean. It would have no experience of individual objects, only of the surrounding water. Motion, temperature and pressure would provide its basic sensory data. In such a pure continuum the discrete would not arise and there would be nothing to count.

This may not be a strong argument, however. For if the jellyfish has consciousness of its individual existence, as in "I am," it should, if endowed with some intelligence, be able to identify the number one. It could then start either playing with this number, say adding or subtracting it to itself, or creating a set with two elements, empty and jellyfish. From that, it can create a new set including this one plus another jellyfish and so on. It seems that any being with consciousness will learn how to count from identifying itself as a unit. (And if it has heartbeats or other periodic functions, it will be easy to do so.)

8. Albert Einstein, "Remarks on Bertrand Russell's Theory of Knowledge," in *The Philosophy of Bertrand Russell*, ed. Paul Arthur Schilpp, Library of Living Philosophers, vol. 5 (Evanston, IL: Northwestern University Press, 1944), 287.

9. Mario Livio, *Is God a Mathematician?* (New York: Simon & Schuster, 2009), 238.

10. Eugene Wigner, "The Unreasonable Effectiveness of Mathematics in the Natural Sciences," *Communications in Pure and Applied Mathematics* 13, no. 1 (February 1960).

11. Hardy, *A Mathematician's Apology*, 37.

12. Benoît Mandelbrot, *The Fractal Geometry of Nature* (New York: Freeman, 1982), 1.

13. I refer the interested reader to my book *A Tear at the Edge of Creation*, in which I tell the story of antimatter in great detail.

CHAPTER 30: INCOMPLETENESS

1. I benefitted tremendously from the classic *Gödel's Proof*, by Ernest Nagel and James R. Newman, and the foreword by Douglas R. Hofstadter, revised ed. (New York: New York University Press, 2002).

2. Hofstadter, foreword to Nagel and Newman, *Gödel's Proof*, xiv.

3. For readers interested in understanding these limitations more deeply, I suggest: Gregory Chaitin, Newton da Costa, and Francisco Antonio Doria, *Gödel's Way: Exploits into an Undecidable World* (London: CRC, 2012). I am proud to note that Francisco Antonio Doria was my master's thesis advisor and collaborator, a wonderful mentor during the early stages of my career. I also suggest Chaitin's *Meta Math!: The Quest for Omega* (New York: Vintage Books, 2005).

4. As we have seen, this is an incorrect interpretation of Einstein's theory of relativity; it does exactly the opposite, providing a clear-cut method for different observers to compare their measurements and resolve any apparent contradictions caused by their mutual relative motion.

CHAPTER 31: SINISTER DREAMS OF TRANSHUMAN MACHINES: OR, THE WORLD AS INFORMATION

1. Nagel and Newman, *Gödel's Proof*, 112.

2. John McCarthy, Marvin Minsky, Nathan Rochester, and Claude Shannon, "A Proposal for the Dartmouth Summer Research Project on Artificial Intelligence," August 31, 1955, 1, http://web.cs.swarthmore.edu/~meeden/cs63/f11/AIproposal.pdf.

3. Ray Kurzweil, *The Singularity Is Near: When Humans Transcend Biology* (New York: Viking, 2005).

4. Second place goes to IBM's Sequoia, with 16.32 petaflops and over 1.5 million core processors.

5. Noam Chomsky, *Language and Problems of Knowledge* (Cambridge, MA: MIT Press, 1988), 152. Similar ideas on the limitations of our

cognitive capacity appear also in Chomsky's *Reflections on Language* (New York: Pantheon Books, 1975), and in Jerry Fodor's *The Modularity of the Mind* (Cambridge, MA: MIT Press, 1983).

6. Thomas Nagel, "What Is It Like to Be a Bat?" in *Mortal Questions* (Cambridge: Cambridge University Press, 1979).

7. From "Tyndall Blogged: Freud's Friends and Enemies One Hundred Years Later, Part 1," *Transcribing Tyndall: Letters of a Victorian Scientist* (blog), February 6, 2010, http://transcribingtyndall.wordpress. com/2010/02/.

8. Colin McGinn, "What Can Your Neurons Tell You?" *New York Review of Books* 60, no. 12 (July 2013): 50.

9. David Chalmers, "Facing Up to the Problem of Consciousness," *Journal of Consciousness Studies* 2 no. 3 (1995): 200–219.

10. In particular, I note philosopher Patricia Churchland's latest book, *Touching a Nerve: The Self as Brain* (New York: Norton, 2013), in which she argues that our current state of ignorance about the brain in no way restricts what can be known in the future. Although I share her enthusiasm for the advancement of learning, I can't agree with her overconfident reason-conquers-all take on human knowledge, for reasons I hope to have made clear in this book. There is humility, and not arrogance, in accepting that we can't know all.

11. Max Tegmak, "The Importance of Quantum Decoherence in Brain Processes," *Physical Review E* 61 (1999): 4194–4206, http://xxx. lanl.gov/quant-ph/9907009.

12. To these musings on our changing cosmic reality we could add our changing material reality, or how we picture the reality of matter. Both are key aspects of our physical reality, addressed in Parts 1 and 2 of this book, respectively.

13. For example, the acuity of the human eye is roughly 0.3 arc-minutes, or 0.3/60 of a degree or 1/200th of one degree. A 20 x 13.3 inch print viewed at 20 inches would require about 74 megapixels to match the limits of the human eye. Higher resolutions, in terms of more pixels, would be spurious. See Roger N. Clark, "Notes on the Resolution and Other Details of the Human Eye," ClarkVision Photography, November 25, 2009, http://www.clarkvision.com/articles/human-eye/index.html.

14. Nick Bostrom, "Are You Living in a Computer Simulation?" *Philosophical Quarterly* 211 (2003): 245–255.

15. And for movie buffs, I should also include Rainer Fassbinder's *World on a Wire*, from 1973, also with a computer Simulacron, which can simulate reality.

16. Seth Lloyd, "The Computational Universe," in *Information and the Nature of Reality: From Physics to Metaphysics*, ed. Paul Davies and

Niels Henrik Gregersen (Cambridge: Cambridge University Press, 2010), 100.

17. Ibid., 102.

18. Seth Lloyd, "Ultimate Physical Limits to Computation," *Nature* 406 (2000): 1047.

19. Seth Lloyd, "Computational Capacity of the Universe," *Physical Review Letters* 88, no. 23 (2002): 237901–237905. Incidentally, the bound on the number of bits comes from applying the famous holographic principle to the Universe as a whole: the maximum amount of information that can be registered by any physical system including gravitational ones (like stars and black holes) is equal to the area of the system divided by the square of the smallest length that we can consider, the so-called Planck length, of about 10^{-33} cm. This is the length scale that marks the transition between classical and quantum gravity. The "holographic" in the name of the principle comes from the idea that all of the information needed to characterize an object can be encoded in its surface. The topic is fascinating but would lead us astray. For further reading I suggest Leonard Susskind and James Lindesay, *An Introduction to Black Holes, Information and the String Theory Revolution: The Holographic Universe* (Hackensack, NJ: World Scientific, 2005).

20. S. R. Beane, Z. Davoudi, and M. J. Savage, "Constraints on the Universe as a Numerical Simulation," November 9, 2012, http://arXiv:1210.1847. In this case, high-energy cosmic rays, which presumably come equally from all directions of the sky, would appear to come from mostly three directions, the north-south, east-west, up-down directions of the lattice. (More technically, isotropy is compromised.) Even if the paper assumes that our simulators will be using similar techniques to ours, albeit optimized, it is still an interesting exercise to consider what sorts of flaws we may expect in a large-scale simulation that attempts to model the Universe from the bottom up.

21. Paul Cockshott, Lewis M. Mackenzie, and Greg Michaelson, *Computation and Its Limits* (Oxford: Oxford University Press, 2012).

22. This is the third of Clarke's "Three Laws." They can be found in Arthur C. Clarke, "Hazards of Prophecy: The Failure of Imagination," in *Profiles of the Future: An Enquiry into the Limits of the Possible*, rev. ed. (New York: Harper & Row, 1973), 14, 21, 36.

CHAPTER 32: AWE AND MEANING

1. With inspiration from Dylan Thomas's poem "Do Not Go Gentle into That Good Night."

Bibliography

HISTORY OF SCIENCE, RELIGION, AND PHILOSOPHY

Bacon, Roger. *The Mirror of Alchimy: Composed by the Thrice-Famous and Learned Fryer, Roger Bachon*. Edited by Stanton J. Linden. New York: Garland, 1992.

Berlin, Isaiah. *Concepts and Categories: Philosophical Essays*. Edited by Henry Hardy. New York: Viking, 1979.

Bohm, David. *Wholeness and the Implicate Order*. London: Routledge & Kegan Paul, 1980.

Burkert, Walter. *Lore and Science in Ancient Pythagorianism*. Translated by Edwin L. Milnar Jr. Cambridge, MA: Harvard University Press, 1972.

Chomsky, Noam. *Language and Problems of Knowledge: The Managua Lectures*. Cambridge, MA: MIT Press, 1988.

———. *Reflections on Language*. New York: Pantheon Books, 1975.

Copernicus, Nicolas. *On the Revolutions of the Heavenly Spheres*. Translated by Edward Rosen. Baltimore: Johns Hopkins University Press, 1992.

Dampier, William Cecil. *A History of Science and Its Relations with Philosophy and Religion*. 4th ed. Cambridge: Cambridge University Press, 1961.

Davies, Paul, and Niels Henrik Gregersen, eds. *Information and the Nature of Reality: From Physics to Metaphysics*. Cambridge: Cambridge University Press, 2010.

De Fontenelle, Bernard le Bovier. *Conversations on the Plurality of Worlds*. Berkeley: University of California Press, 1990.

Eliade, Mircea. *Images and Symbols: Studies in Religious Symbolism*. New York: Sheed & Ward, 1961.

———. *The Myth of the Eternal Return*. New York: Pantheon Books, 1954.

Epicurus. *The Essential Letters, Principal Doctrines, Vatican Sayings and Fragments*. Edited by Robert M. Baird and Stuart E. Rosenbaum. New York: Random House, 2003.

Fodor, Jerry A. *The Modularity of Mind: An Essay on Faculty Psychology*. Cambridge, MA: MIT Press, 1983.

Gleiser, Marcelo. *The Dancing Universe: From Creation Myths to the Big Bang*. New York: Dutton, 1997.

———. *The Prophet and the Astronomer: Apocalyptic Science and the End of the World*. New York: Norton, 2002.

Graham, Daniel W., ed. and trans. *The Texts of Early Greek Philosophy: The Complete Fragments and Selected Testimonies of the Major Presocratics*. 2 vols. Cambridge: Cambridge University Press, 2010.

Grant, Edward, ed. *A Source Book in Medieval Science*. Cambridge, MA: Harvard University Press, 1974.

Greenblatt, Stephen. *The Swerve: How the World Became Modern*. New York: Norton, 2011.

Gregory, Andrew. *Ancient Greek Cosmogony*. London: Duckworth, 2007.

Hughes, Jonathan. *The Rise of Alchemy in Fourteenth-Century England: Plantagenet Kings and the Search for the Philosopher's Stone*. London: Continuum, 2012.

Kahn, Charles H. *Pythagoras and the Pythagoreans: A Brief History*. Indianapolis: Hackett, 2002.

Kirk, G. S., J. E. Raven, and M. Schofield. *The Presocratic Philosophers: A Critical History with a Selection of Texts*. 2nd ed. Cambridge: Cambridge University Press, 1983.

Koyre, Alexandre. *From the Closed World to the Infinite Universe*. Baltimore: Johns Hopkins University Press, 1957.

Lucretius. *The Nature of Things, Book II*. Translated by A. E. Stallings. 1060; rept., London: Penguin, 2003.

Merkel, Ingrid, and Allen G. Debus, eds. *Hermeticism and the Renaissance: Intellectual History and the Occult in Early Modern Europe*. Washington, DC: Folger Shakespeare Library, 1988.

Nagel, Thomas. *Mortal Questions*. Cambridge: Cambridge University Press, 1979.

Newman, William R., and Anthony Grafton, eds. *Secrets of Nature: Astrology and Alchemy in Early Modern Europe*. Cambridge, MA: MIT Press, 2001.

Newton, Isaac. *Four Letters to Richard Bentley*. In *Newton: Texts, Backgrounds, Commentaries*. Selected and edited by I. Bernard Cohen and Richard S. Westfall. New York: Norton, 1995.

Pascal, Blaise. *Pensées*. Translated by A. J. Krailsheimer. Revised ed. New York: Penguin, 1995.

Plato. *The Dialogues: The Republic, Book VII*. Translated by Benjamin Jowett. Great Books of the Western World, vol. 7. Edited by Mortimer J. Adler. 2nd ed. Chicago: Encyclopaedia Britannica, 1993.

Rovelli, Carlo. *The First Scientist: Anaximander and His Legacy*. Yardley, PA: Westholme, 2011.

Rubenstein, Mary-Jane. *Worlds Without End: The Many Lives of the Multiverse*. New York: Columbia University Press, 2013.

Russell, Bertrand. *Our Knowledge of the External World*. New York: Mentor Books, 1960.

Sagan, Carl. *The Varieties of Scientific Experience: A Personal View of the Search for God*. New York: Penguin, 2006.

Sanford, Anthony J., ed. *The Nature and Limits of Human Understanding: The 2001 Gifford Lectures at the University of Glasgow*. London: T & T Clark, 2003.

Schilpp, Paul Arthur, ed. *The Philosophy of Bertrand Russell*. Library of Living Philosophers, vol. 5. Evanston, IL: Northwestern University Press, 1944.

Shirley, John W. *Thomas Harriot: A Biography*. Oxford: Clarendon, 1983.

Yanofsky, Noson S. *The Outer Limits of Reason: What Science, Mathematics, and Logic Cannot Tell Us*. Cambridge, MA: MIT Press, 2013.

GRAVITATION, COSMOLOGY, PARTICLES

Barrow, John D., and Frank J. Tipler. *The Anthropic Cosmological Principle*. New York: Oxford University Press, 1996.

Bojowald, Martin. *Once Before Time: A Whole Story of the Universe*. New York: Knopf, 2010.

Carroll, Sean. *From Eternity to Here: The Quest for the Ultimate Theory of Time*. New York: Penguin, 2010.

Copernicus, Nicolaus. *On the Revolutions of the Heavenly Spheres*. Translated by Charles Glenn Wallis. Amherst, NY: Prometheus Books, 1995.

Davies, Paul. *Cosmic Jackpot: Why Our Universe Is Just Right for Life*. New York: Houghton Mifflin, 2007.

Einstein, Albert. *The Principle of Relativity: A Collection of Original Papers on the Special and General Theories of Relativity*. Translated by W. Perrett and G. B. Jeffery. New York: Dover, 1952.

Frank, Adam. *About Time: Cosmology and Culture at the Twilight of the Big Bang*. New York: Free Press, 2011.

Gleiser, Marcelo. *A Tear at the Edge of Creation: A Radical New Vision for Life in an Imperfect Universe*. New York: Free Press, 2010. Paperback edition: Hanover, NH: University Press of New England, 2013.

Greene, Brian. *The Elegant Universe: Superstrings, Hidden Dimensions, and the Quest for the Ultimate Theory*. New York: Norton, 1999.

Guth, Alan. *The Inflationary Universe: The Quest for a New Theory of Cosmic Origins.* Reading, MA: Addison-Wesley, 1997.

Hawking, Stephen. *A Brief History of Time: From the Big Bang to Black Holes.* New York: Bantam Books, 1988.

Isham, C. J., R. Penrose, and D. W. Sciama, eds. *Quantum Gravity: An Oxford Symposium.* Oxford: Clarendon, 1975.

Kaku, Michio. *Beyond Einstein: The Cosmic Quest for the Theory of the Universe.* New York: Anchor, 1995.

Krauss, Lawrence. *A Universe from Nothing: Why There Is Something Rather Than Nothing.* New York: Free Press, 2012.

Levin, Janna. *How the Universe Got Its Spots: Diary of a Finite Time in a Finite Space.* New York: Anchor Books, 2003.

Magueijo, João. *Faster Than the Speed of Light: The Story of a Scientific Speculation.* Cambridge, MA: Perseus Books, 2003.

Newton, Isaac. *The Principia: Mathematical Principles of Natural Philosophy.* Translated by I. Bernard Cohen and Anne Whitman. Berkeley: University of California Press, 1999.

Randall, Lisa. *Warped Passages: Unraveling the Mysteries of the Universe's Hidden Dimensions.* New York: Harper Perennial, 2005.

Rees, Martin. *Before the Beginning: Our Universe and Others.* New York: Perseus, 1997.

Smolin, Lee. *The Trouble with Physics: The Rise of String Theory, the Fall of a Science, and What Comes Next.* New York: Houghton Mifflin Harcourt, 2006.

Susskind, Leonard. *The Cosmic Landscape: String Theory and the Illusion of Intelligent Design.* New York: Little, Brown, 2006.

———, and James Lindesay. *An Introduction to Black Holes, Information and the String Theory Revolution: The Holographic Universe.* Hackensack, NJ: World Scientific, 2005.

Vilenkin, Alexander. *Many Worlds in One: The Search for Other Universes.* New York: Hill & Wang, 2008.

Weinberg, Steven. *Dreams of a Final Theory: The Search for the Fundamental Laws of Nature.* New York: Pantheon Books, 1993.

Wilczek, Frank. *The Lightness of Being: Mass, Ether, and the Unification of Forces.* New York: Basic Books, 2008.

———, and Betsy Devine. *Longing for the Harmonies: Themes and Variations from Modern Physics.* New York: Norton, 1988.

Woit, Peter. *Not Even Wrong: The Failure of String Theory and the Continuing Challenge to Unify the Laws of Physics.* London: Jonathan Cape, 2006.

QUANTUM PHYSICS

Aaronson, Scott. *Quantum Computing Since Democritus*. Cambridge: Cambridge University Press, 2013.

Albert, David Z. *Quantum Mechanics and Experience*. Cambridge, MA: Harvard University Press, 1992.

Bell, J. S. *Speakable and Unspeakable in Quantum Mechanics: Collected Papers on Quantum Philosophy*. Cambridge: Cambridge University Press, 1987.

Bohm, David. *Quantum Theory*. New York: Dover, 1989.

Born, Max. *The Born-Einstein Letterss: Correspondence Between Albert Einstein and Max and Hedwig Born from 1916–1955, with Commentaries by Max Born*. Translated by Irene Born. London: Macmillan, 1971.

Deutsch, David. *The Beginning of Infinity: Explanations that Transform the World*. New York: Penguin, 2011.

Einstein, Albert. *On a Heuristic Point of View About the Creation and Conversion of Light*. In *The Old Quantum Theory: Selected Readings in Physics*, by D. ter Haar. New York: Pergamon, 1967.

Jammer, Max. *The Philosophy of Quantum Mechanics: The Interpretations of Quantum Mechanics in Historical Perspective*. New York: Wiley, 1974.

Kafatos, Menas, ed. *Bell's Theorem, Quantum Theory and Conceptions of the Universe*. Dordrecht, Netherlands: Kluwer Academic, 1989.

Kaiser, David. *How the Hippies Saved Physics: Science, Counterculture, and the Quantum Revival*. New York: Norton, 2011.

Lederman, Leon M., and Christopher T. Hill. *Quantum Physics for Poets*. Amherst, NY: Prometheus Books, 2011.

Lindley, David. *Where Does the Weirdness Go?: Why Quantum Mechanics Is Strange, but Not as Strange as You Think*. New York: Basic Books, 1996.

Lloyd, Seth. *Programming the Universe: A Quantum Computer Scientist Takes on the Cosmos*. New York: Knopf, 2006.

Przibram, K., ed. *Letters on Wave Mechanics: Schrödinger, Planck, Einstein, Lorentz*. Translated by Martin J. Klein. New York: Philosophical Library, 1967.

Sakurai, J. J., and Jim Napolitano. *Modern Quantum Mechanics*. 2nd ed. Boston: Addison-Wesley, 2011.

Wheeler, John Archibald, and Wojciech Hubert Zurek, eds. *Quantum Theory and Measurement*. Princeton, NJ: Princeton University Press, 1983.

Zeilinger, Anton. *Dance of the Photons: From Einstein to Quantum Teleportation*. New York: Farrar, Straus and Giroux, 2010.

MIND, MATH, COMPUTERS

Chaitin, Gregory. *Meta Math!: The Quest for Omega*. New York: Vintage Books, 2005.

———, Newton da Costa, and Francisco Antonio Doria. *Gödel's Way: Exploits into an Undecidable World*. London: CRC, 2012.

Chomsky, Noam. *Language and Problems of Knowledge: The Managua Lectures*. Cambridge, MA: MIT Press, 1988.

Churchland, Patricia. *Touching a Nerve: The Self as Brain*. New York: Norton, 2013.

Clarke, Arthur C. *Profiles of the Future: An Enquiry into the Limits of the Possible*. Revised ed. New York: Harper & Row, 1973.

Cockshott, Paul, Lewis M. Mackenzie, and Greg Michaelson. *Computation and Its Limits*. Oxford: Oxford University Press, 2012.

Goldstein, Rebecca. *Incompleteness*: *The Proof and Paradox of Kurt Gödel*. New York: Norton, 2005.

Hardy, G. H. *A Mathematician's Apology*. Cambridge: Cambridge University Press, 1940.

Hofstadter, Douglas R. *Gödel, Escher, Bach: An Eternal Golden Braid*. New York: Basic Books, 1979.

Kurzweil, Ray. *The Singularity Is Near: When Humans Transcend Biology*. New York: Viking, 2005.

Lakoff, George, and Rafael E. Núñez. *Where Mathematics Comes From: How the Embodied Mind Brings Mathematics into Being*. New York: Basic Books, 2000.

Livio, Mario. *Is God a Mathematician?* New York: Simon & Schuster, 2009.

Mandelbrot, Benoît. *The Fractal Geometry of Nature*. New York: Freeman, 1982.

Nagel, Ernest, and James R. Newman. *Gödel's Proof.* Revised ed. New York: New York University Press, 2002.

Noë, Alva. *Out of Our Heads: When You Are Not Your Brain, and Other Lessons from the Biology of Consciousness*. New York: Hill & Wang, 2009.

Randi, James. *The Truth About Uri Geller*. Revised ed. New York: Prometheus Books, 1982.

Index

Beane, Silas, 276

Becher, Johann Joachim, 153

The Beginning of Infinity (Deutsch), 227

Being and becoming, atomism and, 138–140

Belief, 3–7
control over Nature and, 9–10

Bell, John, 208–209, 210–213, 217, 222, 230

Bell's theorem, 210–213

Bentley, Richard, 5, 7, 53–54, 57

Berkeley, George, 181, 191, 221–222

Berlin, Isaiah, 17

Bernouilli, Daniel, 153

Big Bang, 30, 84, 87, 97, 112, 113, 119, 132

Black holes, 97, 126

Blue light, 157–158

Blueshift, 68

Bohm, David, 209, 217
complementarity and, 195–196
hidden-variable theory, 185–187, 207–208

Bohr, Niels, 173
behavior of electrons in atoms, 169–170
complementarity and, 195, 196, 199
matrix mechanics and, 171–172
unknowableness of essence of reality, 208

Bojowald, Martin, 98

Borges, Jorge Luis, 227

Born, Max, 171–172, 173, 182–184

Bostrom, Nick, 272, 273, 274

Bosveld, Jane, 148

Boundary conditions, 232

Bousso, Raphael, 120–121, 122

Boyle, Robert, 53, 140, 147–148, 149, 152, 153

Boyle's law, 153

Bracciolini, Poggio, 140

Brahe, Tycho, 36, 37, 39–42
instruments, 37, 39
Kepler and, 42–43, 44

Brain
consciousness and, 266–270
development of abstract conceptual tools and, 250
exaflop machines mimicking, 265
human body and, 262–263
perception of reality and, xv–xvi, 73–76, 192, 218, 235–236, 244, 247, 251, 257, 271–272
quantum effects and, 236
simulation of, 264–265
See also Artificial intelligence (AI)

Brane concept, 119

Bronze, 143

Bronze Age, 143

Bruno, Giordano, 47–48

Buckyballs, 200

Buddha (Siddhartha Gautama), 15

Buridan, Jean, 46

Byron, George Gordon (Lord), 91–92

Caloric, 154–155

"Can [the] Quantum-Mechanical Description of Physical Reality Be Considered Complete?" (Einstein, Podolsky, Rosen), 193–197

Cantor, Georg, 94

Capra, Fritjof, 217–218

Carter, Brandon, 121–122

Cassiopeia, 40

Causality, 225–226

Cause and effect, law of, 44

Centaurus, 82

Chaitin, Gregory, 245

Chalmers, David, 268

Change
atomism and, 138
sensing, 22
time and, 76–77